D0046060

THE MAN WHO FOUND TIME

THE MAN WHO FOUND TIME

JAMES HUTTON AND THE DISCOVERY OF THE EARTH'S ANTIQUITY

JACK REPCHECK

PERSEUS
PUBLISHING

A Member of the Perseus Books Group

Library of Congress Control Number: 2003103400
ISBN 0-7382-0692-X

Perseus Publishing is a member of the Perseus Books Group.
Find us on the World Wide Web at http://www.perseuspublishing.com.
Perseus Publishing books are available at special discounts for bulk purchases in the U.S. by corporations, institutions, and other organizations. For more information, please contact the Special Markets Department at the Perseus Books Group, 11 Cambridge Center, Cambridge, MA 02142, or call (800) 255-1514 or (617) 252-5298, or e-mail j.mccrary@perseusbooks.com.

Set in 11.5-point Bulmer MT by the Perseus Books Group

First printing, May 2003

1 2 3 4 5 6 7 8 9 10—06 05 04 03

For Claire and Jack Repcheck, my parents

CONTENTS

Prologue 1

1 • Looking So Far Into the Abyss of Time 13

2 • First Came Adam and Eve, Then Came Cain
and Abel . . . 25

3 • Auld Reekie 45

4 • The Storm Before the Calm 67

5 • Youthful Wanderings 85

6 • The Paradox of the Soil 103

7 • The Athens of the North 117

8 • The Eureka Moments 145

9 • Hutton's Boswell's 163

10 • The Huttonian Revolution 179

Epilogue 199

Appendix 209

Sources and Suggested Readings 217

Acknowledgments 229

Index 235

PROLOGUE

He burst the boundaries of time, thereby establishing geology's most
distinctive and transforming contribution to human thought—Deep Time.

Stephen Jay Gould, 1977

BEFORE THERE WAS SCIENCE, there was the Bible. For thousands
of years, it supplied reassuring answers to those profound questions that humans have always asked. Who are we? Where are we
in relation to everything else in the universe? And how and when
did we get *here,* this place we call Earth?

The Bible's teachings about the mysteries of existence were
comforting. The Book of Genesis said that an all-knowing, all-
powerful God "created the heaven and the earth" on the First
Day. Over the next five days, all creatures that walk, crawl, and
swim were given life. And God was pleased. Even more soothing
were its teachings about us. Man, we were told, was formed in
nothing less than God's own image, the earth a special home for
His highest creation. Adam and Eve and their descendants had
certainly fouled things up, to the point where God eventually felt

compelled to start over with Noah's family. But, still, man was uniquely blessed, and Earth was his dominion.

One of the few mysteries not resolved explicitly in the Bible was the age of the universe. But learned scribes, teasing information from the Holy Scriptures, and paying close attention to the Hebrew prophesies, had stepped in to supply the answer. They calculated that Creation had occurred not quite 6,000 years ago.

Yet the reverence accorded to biblical answers caused problems, the most serious being that it prevented rigorous and systematic examination of the very world that God had created. Scholars who investigated fields that did not touch on church doctrine were relatively unaffected, but those who explored the natural world were playing with fire—the figurative fire of controversy, the real fire of the heretic's pyre, and the eternal fire of damnation if the church felt they had stepped too far. It required genuine bravery even to venture into these issues; it required hard-to-imagine resolve to promote a position that conflicted with church teachings.

A surprising number of individuals had this unique form of intellectual courage, but it was largely the work of just four men who shattered the biblically rooted picture of Earth and separated science from theology.

The first was Nicolaus Copernicus. A Catholic cleric living in what was then Prussia, Copernicus argued in 1543 that the sun, not the earth, was the center of the universe. All those wondrous heavenly bodies revolved not around man's home, but a ball of fire in the sky. If the earth was no longer the center of things, was it still special? Why would God choose a place other than the center of the universe as the home for a creation made in His own image? Because Copernicus expected his theory to bring on the wrath of church leaders, he waited until the end of his life to publish it. The cleric was on his deathbed when the first copies of his book, *De Revolutionibus Orbium Coelestium* (1543), arrived from the printer.

Because of a cryptic introduction and the technical nature of the work, Copernicus's book did not have a profound impact immediately. It took Galileo, the first celebrity scientist, to publicize the true meaning of what Copernicus had written. Ninety years after Copernicus's death, Galileo was placed under house arrest by the Inquisition for endorsing the Copernican system in his influential book, *Dialogue Concerning the Two Chief World Systems* (1632).

As troubling to the devout as Galileo's endorsement of Copernicus's sun-centered universe was, it was not as bad as what would come next. After all, the Bible did not actually state that the earth was the center of the universe. That doctrine came from St. Thomas Aquinas, the influential Catholic scholar who lived and wrote during the thirteenth century. He took the idea from the Greek astronomer Ptolemy. However, the Book of Genesis did say that the earth was formed on the First Day of Creation and that Adam was created five days later, a sequence that everyone knew had occurred almost 6,000 years ago. The King James Bible, first published in the seventeenth century, verified this common knowledge by placing specific dates for key events right in the margin. Thus, all English-speaking Christians knew that God had created the earth on October 23, 4004 B.C.

James Hutton, a Scottish natural philosopher, boldly confronted this centuries-old wisdom. Writing in 1788, he formally presented proof that the earth was significantly older than 6,000 years. In fact, its age was incalculable—it could be hundreds of millions of years old, it could be billions. Hutton reached his conclusion about the age of the planet through his revolutionary theory of the earth, which recognized the importance of the glacially slow process of erosion coupled with the dynamic movements of earth's surface caused by intense underground heat.

Most previous scholars who had developed hypotheses about the earth had never questioned the church's teachings. They saw Noah's Flood or the waters of the unformed earth as the explanation for all odd geologic formations, thus allowing the age of the earth to fit within six millennia. After the intellectual revolution started by Sir Isaac Newton in the late 1600s, a group of biblical geologists tried to develop sophisticated theories that used modern science to shoehorn the earth's history into 6,000 years. And though a handful of predecessors had questioned whether the history of the physical earth could fit into such a short time frame—one had even calculated the age of the earth to be 75,000 years—the strictures of the past were hard to overcome. Hutton completely ignored the Bible and the Deluge, and as a result he was able to clearly see what rock formations told him.

Hutton's theory was deeply upsetting on two counts. First, it questioned the veracity of the Bible, and second, it displaced humans from close to the start of time. If the Book of Genesis was correct, man was created only five days after the earth was;

if Hutton was correct, the earth had existed for eons before man came along. So, Copernicus took man away from the divine center of things, and Hutton took him away from the divine beginning of things.

Charles Darwin, writing seventy years after Hutton, took the concept of the divine away from man altogether. Darwin's thesis was that far from having been created miraculously by God, the species *Homo sapiens* was simply descended from an ancestor shared with the common ape. No divine intervention was needed.

Of the four, only Copernicus and perhaps Galileo were Christians—Hutton was a deist, believing strongly in a creator God, and Darwin was an atheist. Still, these men were not bent on battling with their respective churches; they were simply seeking the truth unconstrained by past biases, even those sanctioned by the clergy.

COPERNICUS, GALILEO, AND DARWIN are regarded as the key figures in the freeing of science from the straightjacket of religious orthodoxy. But James Hutton must be counted among them. Biblical scholars had proved generation after generation that the first day of Creation occurred in approximately 4000 B.C. In fact, biblical chronology, as the discipline of precise biblical dating was called, was one of the most rigorous "sciences" of the pre-Renaissance era. Beyond scholars, many of the holiest figures from church history, including the prophet Elijah, St. Augustine, St. Bede, St. Thomas Aquinas, and even Martin Luther, had commented on the age of the earth and all had reached the same conclusion: the earth was nearly 6,000 years old. Alongside the belief in the young earth was the equally powerful belief that the earth would not persist indefinitely—the temporal home of God's highest creation was truly temporary. Soon, Jesus Christ would return to his earthly kingdom to lead the final millennium described in the Book of Revelation, and all existence thereafter would be in the paradise of heaven or the horror of hell. But Hutton saw no termination in sight. He stated, "We find no vestige of a beginning—no prospect of an end." Acceptance of Hutton's theory required a complete rethinking of the Christian worldview.

Moreover, Hutton's influence on Charles Darwin was significant. While aboard the HMS *Beagle* in late 1831, en route to the islands where his theory of evolution would begin to be hatched, Darwin carefully studied a recently published book by Charles Lyell, *The Principles of Geology* (1830). Lyell had rediscovered Hutton's work a generation after it had been forgotten by all but a few scholars. For Darwin, the key insight in Lyell's book was that the earth is profoundly old—geologists now believe that it is 4.6 billion years old—an idea that Lyell properly credits to James Hutton in the first pages of his book.

The ancient age of the earth came as a revelation to Darwin. He had been taught at Cambridge University to trust the teachings of the Book of Genesis, and he was at first highly skeptical of claims that refuted what was so widely believed. However, while exploring St. Jago in the Cape Verde Island chain off the coast of Africa—the first stop the *Beagle* made—he noticed an undisturbed layer of rocks, called a stratum, formed of shells and coral. It was so undisturbed, in fact, that it looked exactly like a living coral reef that had somehow hardened to stone. Such a band of shells and coral was not too unusual, but this one was 30 feet above sea level. The only way the stratum of delicate ocean fossils could have been raised so high was through the gradual uplifting of the land, a process that Lyell, and before him Hutton, had described. Gentle uplifting of that magnitude would have taken eons. The stratum on St. Jago showed Darwin that Lyell and Hutton were right—the earth was ancient.

When Darwin left Plymouth harbor just a few weeks before St. Jago, he was a bright but traditional naturalist, a collector of specimens really. Now he was a scientist, and his theory of evolution by natural selection, the key ingredient of which was time—lots and lots of time—began to take shape. If Darwin had not been jolted by Hutton and Lyell into appreciating the age of the earth, it is arguable that he would not have deduced the theory of evolution. If not Darwin, then surely someone, such as Alfred Wallace (who did, in fact, independently discover evolution by natural selection), would have uncovered it shortly afterward. But would it have had the power that Darwin's still outstanding presentation gave it?

COPERNICUS, GALILEO, AND DARWIN are household names; Hutton is anything but. The goal of this book is to change that by telling the intriguing story of James Hutton and the discovery of the earth's antiquity.

For his science alone, Hutton deserves to be better known. In addition to giving geology, as Stephen Jay Gould stated, its most transforming idea—that the earth was ancient—Hutton devised the first rigorous and unified theory of the earth. His theory posited that the earth was constantly restoring itself. He based this concept on a fundamental cycle: erosion of the present land, followed by the deposition of eroded grains (or dead ocean organisms) on the sea floor, followed by the consolidation of those loose particles into sedimentary rock, followed by the raising of those rocks to form new land, followed by erosion of the new land, followed by a complete repeat of the cycle, over and over again.

Hutton was also the first to recognize the profound importance of subterranean heat, the phenomenon that causes volcanoes, and he argued that it was the key to the uplifting of formerly submerged land. It was a completely original theory. Unlike all previous hypotheses of Earth's workings, there was no call for catastrophes, such as Noah's Flood. All of the earth's history could be understood as the result of the subtle actions of common phenomena, such as rain and waves, simply occurring day after day after day, over a profoundly long time. Hutton's proposition was remarkably prescient and essentially correct. His ideas were the starting point for the modern theory of the earth, which now includes plate tectonics and the role of the ice ages.[1]

Beyond James Hutton's scientific contribution, there are several other reasons to explore his life in detail. The first concerns his milieu. Hutton was an integral part of what is now recognized as one of the most creative periods in intellectual history. Starting

---

[1]See the Appendix for a fuller description of these sophisticated concepts.

around 1750, a small group of academics, amateur scholars, government officials, clergymen, and inventors, all about the same age and all centered around Scotland's capital, Edinburgh, made broad and seminal contributions to Western collective knowledge within essentially one generation. This flowering of philosophical, economic, historical, and especially scientific work is now known as the Scottish Enlightenment. David Hume set standards in Western philosophy and history. Adam Smith developed the field of modern economics. Joseph Black isolated carbon dioxide and was among the founders of modern chemistry. Black's former assistant, James Watt, perfected the practical steam engine, which literally powered the Industrial Revolution. Hutton and many of these great thinkers interacted often, sometimes daily, and there is little doubt that the unique quality of life in Edinburgh in the second half of the eighteenth century served as a catalyst for this explosion of creativity.

The final reason to explore James Hutton's life is that it was simply fascinating. He was a late bloomer who came of age during a watershed period of Scottish history. A jack-of-all-trades, Hutton tried being a lawyer, doctor, and farmer before finally finding his true calling as a scientist. Though he was the last of the great Edinburgh scholars to publish his seminal ideas, he commanded the respect of all the other participants in the Scottish Enlightenment. All who came in contact with him noted his animated personality, his energy, and his good cheer. People were simply drawn to him. As Joseph Black once wrote to James Watt, "I wish I could give you a dose now and then of my friend Hutton's company; it would do you a world of good." Once within his orbit, though, attraction to Hutton's personality gave way to admiration for his clear, rigorous, logical, and obviously original mind. His biographer, colleague, and much younger friend, John Playfair, left this profile of our protagonist:

> To his friends his conversation was inestimable; as great talents, the most perfect candor, and the utmost simplicity of character and manners, all united to stamp a value upon it. . . . The simplicity pervaded his whole conduct; while his manner, which was peculiar, but highly pleasing, displayed a degree of vivacity, hardly ever to be found among men of profound and abstract speculation. His great liveliness, added to the aptness to lose sight of himself, would sometimes lead him into little eccentricities, that formed an amusing contrast with the graver habits of a philosophic life. . . . But it is impossible by words to convey any idea of the effect of his conversation, and of the impression made by so much philosophy, gaiety and humor, accompanied by a manner at once so animated and simple. Things are made known only by comparison, and that which is unique admits of no description.

WHAT FOLLOWS, THEN, IS AN EFFORT to trace, largely through Hutton's life, the forces and ideas that came together to prove that the earth was ancient—not eternal, but unknowably, incomprehensibly old. The story is a rich one, filled with odd characters, strong friendships, a uniquely social city, profoundly original ideas, and spectacular geology. The book will explore the notions that Hutton had to confront—the entrenched belief in the young earth, and the accepted geologic theories that used the young-earth time frame as a starting point. It will also examine the two environments that inspired Hutton—the physical environment of Scotland and the intellectual environment created by the members of the Scottish Enlightenment. The ultimate establishment of the ancient earth was the result of a remarkable partnership—Hutton and his young colleagues John Playfair and James Hall—in which one man so inspired his younger charges that they put their own ambitious careers on hold so that they could protect their mentor's legacy. Finally, the book will detail the profound influence that James Hutton had on two nineteenth-century scientists whose work remains powerfully significant today—Charles Lyell, whose view of geology still provides one of the foundations for the field 175 years later, and Charles Darwin, whose theory of evolution only becomes more important as time goes on.

Looking So Far Into
the Abyss of Time

On us who saw these phenomena for the first time, the impression
made will not easily be forgotten. . . . We often said to ourselves,
What clearer evidence could we have had of the different formation of
these rocks, and of the long interval which separated their formation,
had we actually seen them emerging from the bosom of the deep?

John Playfair, 1805

O<small>N A SUNNY</small> J<small>UNE AFTERNOON IN</small> 1788, three gentlemen from
Edinburgh, along with several farmhands, boarded a boat on a
desolate Scottish beach. After clearing the waves, they began
sailing south along the North Sea coast. The men were in search
of a rock exposure on the battered cliffs that would prove one of
the most stunning claims in the history of science—that the earth
was ancient beyond calculation. In the late eighteenth century, as
in all centuries since the formation of the Christian church, this

was a blasphemous statement. The Scottish Presbyterian Church, the English Anglican Church, the Lutheran Church, and the Catholic Church—indeed, all Christian churches, their clergies, and their followers—believed that the earth was not even 6,000 years old. This belief was a tenet based on rigorous analysis of the Bible and other holy scriptures. It was not just the devout who embraced this belief; most men of science agreed that the earth was young. In fact, the most famous of them all, Isaac Newton, had completed a formal calculation of the age of the earth before he died in 1727, and his influential chronology confirmed that the biblical scholars were right.

The assertion that the earth was *not* so young had been made by one of the gentlemen seated in the boat. Three years earlier, Dr. James Hutton, an amateur naturalist who had spent decades studying minerals and their positions in the earth, had addressed an audience of the Royal Society of Edinburgh, which included some of Europe's most accomplished men of science.[1] He boldly announced that his theory of the earth in dicated that it was immeasurably old. Though he certainly commanded the attention of his audience that day, he did not persuade many. Perhaps no one. Because of his failure to gain adherents to his theory, one that he felt was so obviously true, Hutton had spent the last three years doing focused fieldwork in

---

[1] It should be noted that the term "scientist" was not coined until the 1830s; before then, individuals who studied the sciences called themselves natural philosophers. Because this term is awkward for readers in the twenty-first century, I shall use "scientist" from this point forward.

an effort to find geologic formations that would convince his skeptics. Two of them were with him in the boat.

Hutton was now sixty-two years old, but had the energy and stamina of a much younger man. Bald and very thin, he must have looked a bit like Washington Irving's character Ichabod Crane; attached to his long, thin face was a prominent aquiline nose, and his portrait suggests that he had spindly arms and legs. Acquaintances observed that he tended to dress in the same shade of brown nearly every day, and his plain attire was rather out of fashion. Although Hutton had a medical degree, he never practiced medicine and his private income allowed him to lead a life of leisure. Unmarried, he lived with his sisters in a comfortable house in Edinburgh.

Most who knew the doctor agreed that he was a bit odd, or at least eccentric. But, lovably so. He often seemed to lose sight of himself, making unusual faces and movements when he became excited: "The fire of his expression on such occasions, and the animation of his countenance and manner, are not to be described; they were always seen with great delight by those who could enter into his sentiments, and often with great astonishment by those who could not." This passion was especially apparent when he was engaged in conversation, speaking in his broad Scots accent, about some topic that interested him. And it seemed that nearly every topic did interest him. Despite his quirks, his numerous friends were devoted to him, and they recognized that he was a serious and able scholar.

One of the two skeptics on board was John Playfair. A professor of mathematics at the University of Edinburgh since

1785, he was considered among the cleverest men in Scotland, a nation unusually blessed with distinguished intellects at this time. Handsome for his forty years, and very popular in Edinburgh, he had a square jaw and a full head of dark hair combed forward like that of Julius Caesar. Playfair was a brilliant algebraist, geometer, and astronomer, and in the years ahead he would write what would become the standard text in Euclidian geometry for the next half century. Though he had first met Hutton in 1781, he did not know what to expect from Hutton's 1785 lectures. Playfair was as stunned as everyone else in the room by the assertion that the earth was ancient. As a former Presbyterian minister, he had long ago internalized the Scottish church's teaching about the young earth.

The other skeptic was Sir James Hall. A thin blade of a man, with a pleasant face and small features, the twenty-seven-year-old had inherited a substantial fortune while still a teenager. The coastal estate that served as the home base for this excursion was part of that inheritance. Hall also supplied the boat and the additional hands to help sail it. Though still quite young, Sir James was already an accomplished scientist. His granduncle and guardian, Sir John Pringle, was president of the Royal Society, the most esteemed scientific body in the world and the former intellectual home of Isaac Newton. Hall was particularly fascinated by chemistry and had recently become acquainted with the great Antoine Lavoisier in Paris. Visiting the continent when Hutton gave his controversial lectures, Hall nonetheless read a printed thirty-page abstract of the talks that had found its way to him. His reaction mirrored that of the audience: He rejected it.

To call Playfair and Hall skeptics is a bit misleading. By June 1788, after countless face-to-face discussions with Hutton, they had slowly been persuaded that the doctor's theory was plausible. But at the time of Hutton's explosive lectures, the announcement that the earth was ancient was startling. It would be akin to being told today that the sun is not really the source of the earth's heat and light, or that there actually is complex life on the moon. Though other natural philosophers had intimated that the earth was not as young as six millennia, and one famous French scientist had recently argued that the earth was 75,000 years old, no one had ever gone as far as Hutton—or been so direct.

Also, a rigorous theory of the earth's history and workings that most scientists found compelling already existed. This theory was promulgated in the 1770s by a talented German mineralogist named Abraham Gottlob Werner, who was a professor at a renowned German university. Werner argued that a "universal ocean" had once blanketed the earth, creating all the formations that now existed. This was acceptable to the established religions because the universal ocean harkened back either to Noah's Flood or to Creation itself and the very first passages in the Book of Genesis: "And the spirit of God moved upon the face of the waters," and "God said, Let the waters under the heaven be gathered together unto one place, and let the dry land appear; and it was so." Werner's theory also appealed to the scientific community because it seemed to account for all the visible features on the earth.

Hutton's general theory of the earth's history was the opposite of Werner's, as different as heaven from hell. Werner believed that the mountains, deserts, and farmlands had precipitated out of the receding water of the universal ocean; that is, as the ocean slowly evaporated or was drawn into the earth, the land on which humans now lived formed and was revealed. This process had happened only once. Hutton, in contrast, saw new land springing from *below* the already existing oceans, pushed up by the caldron of extreme heat found within the earth. And he saw it happening over and over, the earth following a constant pattern of erosion followed by new land rising up from below the seas— the planet an efficient land-regenerating machine. Hutton reasoned that this cycle had occurred at least three times during geologic history.

For Hutton there was no need to call upon unseen and unknowable catastrophes from the past, such as the Deluge or the universal ocean, to explain geologic formations; they were all understandable through the knowledge of processes still occurring. The inexorable forces of wind and rain, tides and waves, volcanoes and earthquakes, which the earth still experiences every day, formed the world we inhabit. All that was required, as Hutton stated, was "immense time."

SINCE GIVING THE PUBLIC LECTURES, Hutton had been remarkably successful in finding convincing proof that extreme subterranean heat was an active agent in the formation of the continents. Although this finding alone was significant, it did not necessarily follow that intense heat had led to the raising of new land above the oceans to replace the eroded land of former regions. Hutton needed to find an exposure of rocks that somehow demonstrated his theorized cycle.

Discovering such an outcrop was the quest of the three sailors as they plied the waters of the North Sea. Hutton had chosen to investigate here because he knew that this part of Scotland had two distinct types of surface rock. What was believed to be the older of the two was a smooth, grayish stone that mineralogists labeled "primary micaceous schistus" (today it is called Silurian graywacke)—a type of shale. The other, younger rock, a coarse reddish stone, Hutton called the "secondary sandstone strata" (today it is called the Upper Old Red Sandstone). The doctor was convinced that the two rock groups represented two separate erosion-sedimentation-uplift cycles, and that at some location the younger rock (the coarse red sandstone) must come in contact with, and actually cover, the older rock (the gray smooth stone). There was a chance that the junction of the two formations would be visible on the coast, thanks to the intense erosion inflicted by the pounding winds and water of the North Sea.

The men could conceivably have avoided using the boat, and the attendant risk of the sea, by hiking along the coast. However, it was so rugged—there were ravines to cross, steep rock faces to climb, and hills to circumnavigate—that it would have taken days to see everything they were hoping to see on this one day. Besides, Hutton was too old to conduct the exploration by land.

After leaving the spot where they boarded the boat, the Dunglass Burn beach, they sailed along a jagged coastline. The mild weather and low tide allowed them to sail near the shore, and the early afternoon sun gave maximum exposure to the cliffs on their right. The rocks were from 50 to 70 feet high, grass and moss covering the tops. The relentless pummeling by the North Sea gave the sharp juttings an ominous, almost clawlike shape.

About a half mile from Dunglass Burn, the boat came to the first headland, Reed Point. The explorers rounded the point but could detect no unusual formations from the boat; all that was visible was the dominant primary schistus. Hutton decided not to land, and the boat continued southward. They had to be extra careful along the next stretch of coast because, in addition to the rugged cliffs, the waves broke against large rocks protruding from the sea.

After several hundred yards, the boat skimmed past the next headland, and the men turned their heads to witness a spectacular scene. Pease Bay dug deeply into the coast and was marked by a beautiful sandy beach stretching from end to end, at least 200 yards long. What was striking was not the beach itself, but the rock formation that emerged from it. Rising out of the sand, like a snake, was a beautiful red sandstone outcrop, which seemed to burst out of the beach at a low 20-degree angle. The red rocks grew to form a 50-foot cliff. The formation was covered with a thin coating of grass and moss, but enough had been "cleaned" by the surf that the strata were clearly visible. These were the secondary strata that Hutton was looking for. Still, as beautiful as this exposure was, it did not contain the combination of rock layers that Hutton hoped to find.

The team continued south. As the sailors looked up at the rocky cliffs, they saw the angular stone wall of a now roofless and abandoned chapel at the top of one of the hills, probably just a few hundred yards from the edge of the bluff. It was an unusual sight in such a desolate spot. Maybe this previously sacred land would now mark a different kind of shrine.

The boat quickly neared Siccar Point, the next headland on their course. After they rounded the protrusion, Hutton urged the pilot to land the boat. The sand of Pease Bay unfortunately did not extend to Siccar Point, so the boat scraped to a halt on rutted stone. But no one cared much about the boat. As they stood on the rocky beach staring up at the cliff face to their right, it was as if they were looking at a painting left by the Creator to show the wonder of His world. At the bottom of the cliff was the gray-colored primary micaceous schistus exposure, but the layers were not horizontal like the ones seen on a typical quarry wall. They were vertical, standing straight up, like a row of books on a shelf. Above the booklike layers sat a couple of feet of nondescript muddle, composed of large fragments of the schistus. Then, above the hardened muck was another large exposure of layered rocks, but these layers were horizontal and they had the distinct red hue of the exposure just seen at Pease Bay.

Hutton, an animated man at all times, was gleeful. Upon collecting himself, he explained to his companions what they were observing. The schistus that was now vertical had originally been laid down in horizontal deposits, the only way that sediments can form. Eroded grains from an ancient continent had flowed into a sea and settled at the bottom. Since deposits usually settle at a modest rate, perhaps only an inch a year, it took hundreds of thousands of years for enough sediment to build up and apply the pressure to the bottom layers that caused them to be changed to rock. Subterranean heat also assisted in this transformation.[2] Then, the intensity of the heat, and perhaps some other additional force, had caused the once horizontal strata to buckle the way a leather belt would if you held it taut between your hands and then brought your hands together. As a result, the layers folded and became vertical; in the process they also rose above sea level. The once-submerged stratified rocks had become dry land. Immediately, erosion began to work its magic all over again, causing the removal of the tops of the buckled rocks. Over time, this land became covered by water again, either from the sea level rising or the land sinking, because the layer of stony muddle represented the early stages of submersion, when waves broke up rocks along the shore. Then, as the vertical stratified rocks settled deeply under the water, new sediments started piling up, this time formed with red-hued grains from different surface rocks. Eventually, these new sediments also consolidated into rocks, affected by pressure and the same subterranean heat that had once acted on the vertical strata. Hutton and his friends were now looking at this dry-land exposure because the area had been raised above the sea yet again, but with less violence this time since there was no new buckling. Collectively, the making of Siccar Point must have taken an unfathomable length of time— much, much longer than 6,000 years.

---

[2]Hutton believed that both pressure and heat were needed to form sedimentary rocks; it is now known that heat is not necessary for the formation of sedimentary rocks, though it is needed for metamorphic rocks. See Appendix for details.

Finally, here was irrefutable proof. The earth was immeasurably old. How old? Who could even venture a guess? How old was the sun? How old was the solar system?

John Playfair would later write of this triumphant moment:

> We felt ourselves necessarily carried back to the time when the schistus on which we stood was yet at the bottom of the sea, and when the sandstone before us was only beginning to be deposited in the shape of sand or mud, from the waters of a superincumbent ocean. An epocha still more remote presented itself, when even the most ancient of the rocks instead of standing upright in vertical beds, lay in horizontal planes at the bottom of the sea, and was not yet disturbed by that immeasurable force which has burst asunder the solid pavement of the globe. Revolutions still more remote appeared in the distance of this extraordinary perspective. The mind seemed to grow giddy by looking so far into the abyss of time.

BACK ON THE BOAT AND HEADING north again to the Dunglass Burn beach, the explorers were no doubt aware of the importance of their find. Hutton was such a positive and rational man, so generous in his opinion of others, that he probably underestimated the forces that would rise up against him. But Playfair likely did not. As a former Presbyterian minister, he understood the hold the church's teachings had on people, educated and uneducated. The belief that the earth was less than 6,000 years old was deeply entrenched in the psyche of most Christians. Just

as important, Playfair knew how vigorously the church pro-
tected subjects on which it held a position. The church and the
scholars who supported it would not graciously cede the history
of the earth to the impious, perhaps blasphemous, Hutton. The
battle for the truth was just beginning.

2

First Came Adam and Eve, Then Came Cain and Abel . . .

And Adam lived a hundred and thirty years, and begat a son in his own likeness, after his image; and called his name Seth:

And the days of Adam after he had begotten Seth were eight hundred years: and he begat sons and daughters:

And all the days that Adam lived were nine hundred and thirty years: and he died.

And Seth lived an hundred and five years, and begat Enos:

And Enos lived ninety years, and begat Cainan:

Gen. 5:3–9

THE BISHOPS HUDDLED IN TWOS and threes while seated on the wooden benches that ringed the large ornate room. They were anxious for the ceremony to begin because the heat of the mid-morning was already starting to penetrate the hall. Quietly, the large door at the end of the room opened, and a dozen men dressed simply but impressively in clean white robes silently formed a semicircle. With no fanfare, the Roman Emperor Constantine I finally emerged. Looking like an oriental sultan in a purple robe embellished with a large medallion, Constantine walked slowly past the now-standing bishops. When he reached the front of the gathering, he was brought a small wooden stool veneered with gold. If the bishops were going to sit on the hard benches day after day, then Constantine would join them in their discomfort.

Constantine turned to face the bishops, sat down on his stool, and signaled for the others to do the same. Thus began the Council of Nicaea, the first-ever meeting of bishops intended to formalize Christian doctrine.[1] Constantine spoke in Latin, his native language, but many of those present knew only Greek and needed whispered translations. During the few minutes he spoke, he stressed his appreciation for those who had traveled so far and his determination that the deliberations of this meeting be undertaken with the utmost seriousness.

Constantine had called this extraordinary and unprecedented convocation of bishops in the summer of A.D. 325 because there was a crisis in the church, and he wanted it resolved immediately. The crisis was a deepening controversy

---

[1]Nicaea is now called Iznit, Turkey, and it was chosen by Constantine because it was close to the newly founded and still-under-construction city of Constantinople.

over two views of Jesus Christ. One group of bishops believed that Christ was equal in every way to God the Father. The other believed that because Jesus had been born, it followed that he had not existed forever, and therefore he could not be completely equal to the Father. Constantine had no opinion on the matter. He just wanted the argument ended so that he could go back to converting pagans to Christianity—the task that he saw as his true mission as emperor. If his own religion was riddled with controversy, how could he expect it to be attractive to nonbelievers? So resolute was he about complete attendance that he had ordered the invitations to be hand-delivered by Roman soldiers. The solders then waited to escort the attending bishops to Nicaea.[2]

It was Constantine who had given the Christian bishops the freedom to argue over such questions. Just over a dozen years earlier, Christians had feared for their very existence. But in 312, Constantine saw a vision of a cross that inspired him to win a critical battle. He was sure the vision had been sent by Christ. Constantine immediately converted to Christianity and then issued the Edict of Milan, which legalized the faith. Over the next few years, as Constantine succeeded in unifying Roman territory, Christianity became the official religion of the empire.

Sitting in the circle of bishops listening to Constantine was Eusebius, the bishop of Caesarea. He had endorsed the view that Jesus Christ was not equal to the Father, which placed him

---

[2]Pope Sylvester, who was very frail and elderly at this time, was one of the few church officials not in attendance.

in the minority, and he was effectively on probation by his fellow bishops. The Council of Nicaea would decide whether he was to be excommunicated. Now an old man in his mid-sixties, Eusebius had already spent time in prison as a persecuted Christian when he was a young man in his native Palestine. He surely did not want to return to a prison cell.

Eusebius was a prolific writer who, in earlier years, had written several popular books, including one on the history of the martyrs. One of his strategies for returning to the good graces of his fellow bishops and the church was to make a direct appeal to the emperor. He decided to prepare a new edition of one of his old books in honor of Constantine. The book was a chronology of world history, which he had originally written some thirty years earlier. It was a compendium of all the known peoples and their histories from the beginning of time until the present. He also wrote a completely new book for the emperor, this one a history of the Christian church, starting with the birth of Christ and ending with the conversion of Constantine.

When Eusebius gave the books to Constantine, the emperor was pleased. He was particularly fascinated by how Eusebius had gone about constructing his chronology, combining as he did the histories of the Jews, Egyptians, Assyrians, Greeks, and Romans. In addition to straight narrative, Eusebius had created graphlike grids in which he synchronized the various events of different cultures. Here, for instance, is his narrative description of the birth of Christ:

"It was in the forty-second year of the reign of Augustus and the twenty-eighth after the subjugation of Egypt and the death of

Antony and Cleopatra, with whom the dynasty of the Ptolemies in Egypt came to an end, that our Savior and Lord Jesus Christ was born in Bethlehem of Judea."

And, here is his graphic rendering of the same seminal event:

Year of Abraham	Olympiad	Roman	Egyptian
2010	OL 194	Augustine 42	28

Every entry in the graphs began with the "Year of Abraham"; that is, the number of years after the birth of Abraham. The "Olympiad" column refers to when the event happened according to the Olympiad system, the four-year periods that the Greeks started using in the early eighth century B.C.

Constantine admired the rigor of this work, and he ordered many copies to be prepared and circulated throughout Roman territory. Because of this official support, Eusebius's chronology would soon be known throughout the empire. And it did serve its immediate purpose—the emperor insisted that Eusebius be allowed to retain his position as bishop.

The chronology acquired even greater influence seventy years later when St. Jerome translated Eusebius's Greek into Latin, thus making the book available in the two languages of Christendom. St. Jerome's involvement with the text demonstrates its status; he was ordered by Pope Damascus in 382 to perform two translations for the church, the Bible and Eusebius's *Chronology*. St. Jerome's translation of the Bible, finished in A.D. 405, was called the Vulgate; it would be the first book printed by Johannes Gutenberg over a thousand years later, and it would serve as the basis of the King James Bible. It is still the foundational Christian Bible of all those in use today. Similarly, St. Jerome's translation of Eusebius would be the document that would inspire chronologists for the next 1,400 years, reinforcing the notion of a young earth and making James Hutton's attempt to refute it far more difficult.

IN WRITING HIS CHRONOLOGY, Eusebius borrowed openly and with acknowledgment from three sources: the Hebrew Bible (what essentially became the Old Testament), and the works of two earlier writers whose efforts might have been lost if not for Eusebius—Flavius Josephus and Julius Africanus. The version of the Hebrew Bible used by Eusebius is now called the Septuagint Bible, roughly translated to mean "the seventy." Legend has it that in 286 B.C., Ptolemy II, whose father had succeeded Alexander the Great in the newly established Egyptian kingdom, ordered the book of the Jews to be translated into Greek so that he would gain the favor of his recently conquered Jewish population. To complete this task correctly and quickly, he brought seventy-two Israelites—six from each of the twelve tribes—to the capital city of Alexandria, and he gave them seventy-two days to finish their work. The scholars were inspired by God, and though they worked in isolation from one another, their efforts matched perfectly when finished. The truth is quite different. When Ptolemy II came to power, a standard version of the Hebrew Bible did not exist; the Holy Scriptures varied from tribe to tribe and were largely based on oral tradition. That tradition was being threatened by the disuse of Hebrew among the Jews in Egypt. To impress the large Jewish population in his kingdom, Ptolemy II sponsored a translation, and during his long reign scholars working in the great library of Alexandria produced one.

This version of the Hebrew Bible had tremendous influence. Since it was in Greek, then the dominant language of the Middle East, it was accessible to all Jews. Perhaps more important, because the book was conceived and written in the thriving intellectual capital of Alexandria and sanctioned by the emperor, many copies were produced and distributed. In fact, it was one of the most copied books in the world before the invention of the printing press. This was the Bible that Jesus and all his followers used.

For Eusebius's chronology, the Septuagint Bible was important because the first part, the Pentateuch, also known as the Book of Moses and containing the Book of Genesis, stressed the history of the Jews, from Creation to the death of Moses. It gave explicit years for the life spans of the various individuals mentioned, so that every Jew knew Adam had lived to be 930 years old, Noah 950, and that Abraham was 175 and Moses 120 when they died. For Eusebius and all future chronologists, these explicit life spans were always the starting point.

Three and one-half centuries after the death of Ptolemy II, the next key historical manuscript was created. Unlike the Septuagint Bible, this one had but one author, Flavius Josephus. Josephus was born in Jerusalem in A.D. 37, just a few years after the death of Christ. He was one of the leaders of the failed Jewish revolt against the Romans in A.D. 66. After the rebellion was squashed, Josephus went into hiding. He mysteriously survived a suicide pact with the other rebel leaders, only to be captured by Roman soldiers and taken to Rome. He must have been quite charismatic because he convinced Emperor Vespasian to let him live as a free man in Rome. And live he did. For the rest of his life this former rebel led a privileged life among the Roman elite. Josephus became an extraordinarily prolific writer. His most important work was a history of the Jews, which he titled *Jewish Antiquities.* Somewhat ironically, given that Josephus owed his life and high status to his Roman supporters, the goal of *Antiquities* was to demonstrate that the Jews had the longest history of all peoples, including the Greeks and Romans. Using the Septuagint Bible and later works of Jewish scholars, his history began with Creation and ended with the Jewish war against Rome, the one he helped lead.

Josephus's *Antiquities* was the first attempt at a world history, but he was not concerned about determining the age of the world. That distinction fell on Julius Africanus, the first true chronologist.[3] Though Josephus had used Greek and Roman historians as sources for the last centuries of his history, he had depended exclusively on the Bible for the previous millennia.

---

[3]Julius Africanus was the father of biblical chronology and his work was long and elaborate, but Theophilus of Antioch (115–180) produced a brief yet detailed chronology about fifty years before him (it was found in the eleventh century); it is not known whether Julius used Theophilus's work.

Not Julius Africanus. Julius's history of the world, which he titled the *Chronologia*, was the first attempt to juxtapose Hebrew sources with ancient Greek, Egyptian, and Persian sources to locate dates, an innovation that Eusebius would take to a new level. Look, for example, at how Julius handled Moses: "And if one carries the calculation backwards from the end of the captivity, there are 1,237 years. Thus, by analysis, the same period is found to be the first year of the Exodus of Israel under Moses from Egypt, as from the 55th Olympiad to Ogygus, who founded Eleusis. And from this point we get a more notable beginning for Attic chronography."

Julius was born in what is now Libya in about A.D. 160. He was a pagan who served in the Roman army and then discovered Christianity. By 215 he was ordained, had studied at the academy in Alexandria, and was the priest for the town of Emmaus. He wrote his chronology between the years 212 and 221.

How Julius constructed his chronology is worth looking at in detail because his approach was essentially followed over and over again by all future chronologists. Though Eusebius's chronology would be more widely read, it relied heavily on Julius for basic information and techniques. The Old Testament was Julius's key document, the major events being the benchmarks for the chronology. The Old Testament is largely a history of the Jews. It begins with Creation, and continues with the stories of Adam and Eve, Noah and the Flood, Abraham and the entry into the Promised Land, the stories of Abraham's son Isaac, Isaac's son Jacob, and Jacob's son Joseph. The first part ends with the enslavement of the Jews by the Egyptians, and then Moses leading the Exodus of the Jews out of Egypt and the delivery of the Ten Commandments. After the death of Moses, the history of the Jews becomes more convoluted. The highlights that Julius focused on included Joshua and the capture of Jericho, the reign of King David (the David who slew Goliath), King Solomon and the building of the great Jewish temple in Jerusalem, the fall of Jerusalem to the Babylonians, the Jewish captivity in Babylonia, and finally the Jews' triumphant return to Jerusalem and the rebuilding of the temple.

Julius's first task was to determine how much time had elapsed from the birth of Adam, which was five days after the beginning of Creation, to the Deluge, or Noah's Flood. Using the ages of the descendants of Adam detailed in the Septuagint, Julius determined that the rains started 2,261 years after Creation. He believed that the Flood lasted twelve months, thus the year 2262 marks the beginning of the post-Flood period.

The next chapter in the *Chronologia* covers the period from when Noah first stepped off the ark to when the great father of the Jews, Abraham, entered the Promised Land. Julius calculated that this period lasted another 1,015 years. So Abraham crossed the Euphrates River into Canaan in the year 3277. By Julius's reckoning, Abraham represented the twentieth generation after Adam.

The third major chapter is from the entry into the Promised Land to when Moses delivered the Ten Commandments. Julius, using various sources in addition to the Bible now, arrived at a period of 430 years. Thus, the Exodus of the Jews from Egypt under Moses's leadership occurred in the year 3707.

From Exodus on, Julius's chore became more difficult because the Book of Moses, which had paid such close attention to the ages of the Hebrew forefathers, comes to an end. Not to be deterred, Julius calculated that 585 years separated the Ten Commandments and the dedication of the great Jewish temple in Jerusalem, built by King Solomon, bringing the chronology up to the year 4292. The period beginning with the dedication of the temple and culminating in its destruction by the Babylonians lasted another 651 years. Finally, the birth of Jesus Christ took place in the 5,500th year after Creation.

JULIUS BROUGHT HIS *Chronologia* up to A.D. 221, the year he completed the book. He was explicit in stating that Jesus Christ was born five and one-half millennia after the beginning of time. This was significant, because Julius was not simply writing a world history with a focus on dates. In fact, his real purpose was to give context to biblical prophesy. He was most concerned with predicting the second coming of Christ, the thousand-year reign described in the Book of Revelation, the last book of the New Testament. This would be another characteristic of all future chronologies; despite the rigor and attention to detail, their ultimate purpose was to determine when the temporal world of evil and suffering would end and the profoundly better world led by Jesus would begin. Early Christians were persecuted and their savior had been murdered, thus the Second Coming was not just an intriguing idea; it was an anxiously awaited event. Julius predicted that the present world would continue until the year A.D. 500—6,000 years after Creation.

After that, the Messiah—Jesus Christ—would return to the terrestrial sphere to begin his thousand-year reign, the period of heaven on earth. All future chronologists would calculate the earth's age to be 6,000 years at the time of the Second Coming. In the year 1650, the last well-known chronologist, James Ussher, would date the beginning of the world at 4004 B.C.; this gave him nearly 350 years until the end of the sixth millennium.

Where does the 6,000 years come from? Julius was merely the first to put into writing what was a long-held popular belief. This belief came from a conflation of various revered nonbiblical writings and specific passages from the Bible. Perhaps the most important passage from Scriptures came from the Jewish Talmud. There was found the famous prophesy ascribed to the prophet Elijah that stated the world would exist for 6,000 years, the first 2,000 being the void, the next 2,000 being the period of the Torah, and the last 2,000 being the period of the Messiah. Equally influential was a collective exercise in popular numerology. The number 7 is found throughout the Book of Revelation—it was considered a number that represented God—and it coincided with the seven days of Creation in the Book of Genesis. But on the last day God had rested. Thus, it actually took only six days to create the earth and all the creatures that lived on it. The first six days of Creation coincide with the number 6 in Revelation: the number for Satan.

All Christians and Jews knew that the world became contaminated the day Satan tempted Eve, causing Adam and Eve's ejection from the Garden of Eden. Only with the Second Coming would Satan be destroyed and evil eliminated. Revelation clearly stated that after the Second Coming, Christ would rule for a millennium. Two specific places in the Bible state that a day is like 1,000 years to God: In the Old Testament, Ps. 90:4 states, "For a thousand years in thy sight are but as yesterday when it is past, and as a watch in the night"; and, in the New Testament, 2 Pet. 3:8 states, "But, beloved, be not ignorant of this one thing, that one day is with the Lord as a thousand years, and a thousand years as one day." The popular belief was that the coming thousand-year reign of Jesus Christ must be the seventh millennium, and that Satan had dominated the previous six millennia. In many sources, in fact, the final millennium was called "the great Sabbatism"—the great Sabbath, drawing an even more solid connection to the seventh day of Genesis. The first coming of Christ was during the sixth millennium, coinciding with the sixth day of Creation, the day Adam was created.

Julius's great contribution was to give Christians the documentation and proof they wanted for an idea that was already fully embraced. Eusebius, by being the church's first key historian at the time of Constantine and the establishment of the legal Christian church, provided the needed ratification of Julius's chronology. Julius's and Eusebius's careful works would be the benchmarks for all future chronologists; their successors would simply dot i's and cross t's. *But*, their successors would continually push back the end of the 6,000 years, as each threshold for the Second Coming neared. St. Jerome, Eusebius's translator, was the first to practice this form of recalculation; he placed the birth of Christ at 5,200 years since Creation, putting off the end of the sixth millennium until A.D. 800. This kind of fudging was easily done and sparked no controversy because there was enough uncertainty in the original figures to allow for reinter-pretation. The chronologists were consistent in putting off the end of the sixth millennium until a couple of hundred years after their own deaths.

AT THE TIME THAT ST. JEROME was finishing his translation of Eusebius into Latin, the Roman Empire was teetering on its last legs. Less than 100 years earlier, when Constantine had consoli-dated the Roman Empire and made Christianity its official reli-gion, the empire was as dominant as ever. Unfortunately for Rome, its enemies were numerous, they were arrayed all along the extensive border, and they had learned from their strategic mis-steps over the past 700 years. When they attacked Rome now, they struck with gigantic armies, often numbering well over 100,000 men. A series of bloody conflicts in the first decade of the fifth cen-tury culminated in 410 with the sacking of the city of Rome by the overwhelming Visigoth army, led by Alaric. This marked the first time Rome had fallen to invaders in over a thousand years. It also marked the beginning of the end of the Roman Empire in Europe and the commencement of what is now called the Dark Ages.

Of course, the Dark Ages did not begin on August 24, 410, the day that Rome capitulated. The disintegration of the Roman Empire was a gradual process. In the west, though the sacking of Rome was a major turning point, it took several decades before the Visigoths, Vandals, Burgundians, and Franks wrested com-plete control from Rome. By 476, though, it was over. After Attila the Hun had invaded the Italian peninsula and the Vandals had sacked Rome a second time, Italy and all of Europe were controlled by the various Germanic and Baltic tribes. In the east, the old Roman Empire held on for another couple of centuries in the guise of the Byzantine Empire, but it was eventually overrun by the Bulgars from the northwest and the Muslims from the east.

Though the political structure of the empire collapsed slowly, Roman culture unraveled quickly. The loss of stability in the early fifth century took a high toll on scholarship and education. Though the tribes that overran the Roman Empire were not the bloodthirsty barbarians of popular myth, their cultures were martial and violent. Because they had no written languages, there was no legacy of scholarship. Thus, the traditions that cultivated Josephus, Julius, and Eusebius eventually disappeared. For the next thousand years, learning in Europe would be largely confined to the Catholic monasteries and seminaries; these had sprung up in Roman and "barbarian" territory in the fourth and fifth centuries and were left unmolested by the Visigoths, Vandals, and Franks.

Because of the general chaos and uncertainty that followed the Roman Empire's collapse, it is not surprising that apocalyptic prophesy grew increasingly popular. Just as Hebrew prophesy had its greatest flowering during the decades the Jews were enslaved by the Babylonians, the centuries after the fall of Rome represented another great flowering. Many new chronologies were produced, most very explicit about predicting the end of the sixth millennium and the Second Coming. They retained Julius's basic scheme, but they adjusted the end dates, and they often drew an even tighter connection to the six days of Creation.

The most influential chronologies were calculated by Isidore of Seville (560–636) and Bede the Venerable (673–735). Later chronologists became more daring. Joachim of Fiore's (1135–1202) writings drove thousands of peasants in southern Europe into mass hysteria as they waited for the Second Coming in 1260. One of the more intriguing chronologists, Otto of Freising (1111–1158), argued that the thousand-year "binding" of Satan (an interpretation of Revelation) was occurring at precisely that moment. Freising asserted that the Devil had been bound in 325 by the Council of Nicaea, which had unified the church, and he would remain bound until 1325. Afterward, he would escape and wreak havoc for a short period before the Second Coming of the Savior.

THE MEDIEVAL CATHOLIC CHURCH may have preferred the unbinding of Satan to what actually happened less than 200 years later. It is difficult to exaggerate the influence of Martin Luther, so successful was his reformist movement, launched with the posting of the 95 Theses on the door of Wittenburg Cathedral in 1517. The speed and extent of the Protestant Reformation in Europe were unprecedented. Prophesy became pronounced again as new Protestants looked for signs in the Scriptures that predicted their breaking from the church. Many prophetic writings were circulated, invariably with the pope cast as the antichrist. Luther himself produced many of these antipapal apocalyptic tracts.

For a variety of complicated reasons, Luther stopped active participation in the administration of his new church rather early,

but he never stopped teaching and preaching. And, most important, he never stopped writing. His followers waited eagerly for
his words, which were printed and distributed widely (Luther
was the first religious reformer to have access to the printing
press, invented in the 1450s). Luther was strident in his belief
that the Bible was the inspired word of God, and that the Scriptures must be interpreted literally. He turned to chronology late
in life, when he wrote *Supputatio Annorum Mundi.* As with
everything he published, his chronology had a tremendous
impact. He followed the basic style of Julius and Eusebius, but he
made a significant adjustment for the end of the sixth millennium. Writing in 1541, Luther calculated that Creation occurred
in 3961 B.C., thus giving his Protestant followers nearly 500 years
to prepare for the return of Christ. From this point on, chronologies would follow Luther's lead and consistently set Creation at
approximately 4000 B.C. To justify his sharp adjustment, Luther
leaned heavily on the prophesy of Elijah, which proclaimed that
the last 2,000 years of the total 6,000 would be the Age of the
Messiah. Thus, the Second Coming would occur 2,000 years
after the birth of Christ, around the year 2000. In fact, on the very
title page of his chronology, Luther quoted the famous prophesy:

> Elia Propheta, Sex milibus annorum stabit mundus.
> Duobos milibus inane. Duobus milibus Lex.
> Duobus milibus Messiah. Insti sunt Sex dies hebdomadae
>   coram Deo.
> Septimus dies Sabbatum acternum est. Psalm 90. Et 1 Pet 2
> Mille anni sicut dies unus.

A century after Luther, the last of the influential chronologists, James Ussher, archbishop of Armaugh, refined Luther's
calculations. More than 2,000 printed pages, the *Annales Veteris
Testamenti (Annals of the Old Testament;* 1650) took the Calvin-
ist Ussher over two decades to complete. *Annales* left a lasting
legacy because it was used by the editors of the King James
Bible, where Ussher's dates were placed in the margins. After
biblical chronology fell out of fashion in the nineteenth century,
the archbishop became infamous for the outlandishness of his
precise statements, the most familiar being that Creation started
on Sunday at noon on October 23, 4004 B.C. However, the precision found in the *Annales* was very much in keeping with the
tradition of chronology as it had evolved over the centuries. The
King James Bible was produced in vast numbers and it was in-
expensive enough for every Christian family in the English-
speaking world to own one; thus Ussher's calculations were
ingrained in the minds of several generations of British subjects
before James Hutton announced his findings.

The *Annales* would be the high-water mark for biblical
chronology, the final chapter of a tradition that began with the
Septuagint Bible, continued with Julius Africanus and then
Eusebius, was passed on to Isidore and Bede, and then Luther.
A modern reader might well ask, "Surely the scientific community didn't blindly follow such religious documents?" As an
indication of just how difficult James Hutton's task was, the
most famous scientist of his era, Isaac Newton (1642–1727),
also privately produced his own chronology toward the end
of his life. Moreover, Newton was following in the footsteps of

the most important astronomer before him, Johannes Kepler (1571–1630), whose chronology built on Luther's and whose calculations were, in turn, used by Ussher. Kepler, in fact, is credited with the discovery, now confirmed, that Christ was born not in the year 0, but rather in 4 B.C. Newton was a surprisingly spiritual man who was captivated by the Book of Revelation and the Book of Daniel, which is the most prophesy-filled chapter of the Old Testament. Newton admitted to a princess in 1716 that he had been working on a chronology, off and on, for decades. His secret out, Newton exerted a great deal of energy in the last years of his life to keep the chronology from the public. But after he died in 1727, his chronology was quickly disseminated. Like all those that had come before, it essentially followed Julius's outline, though the specific dates largely agreed with Ussher's.

In June 1726, Isaac Newton was an eighty-three-year-old man passing the last year of his life in the village of Kensington, not far from London. Three hundred miles to the north, a baby was born to William and Sarah Hutton in the city of Edinburgh. The child, James, would one day extend Newton's discovery that nature obeyed constant laws, putting an end to 1,500 years of biblical chronology, and profoundly changing the way Christians viewed the world.

3
=

Auld Reekie

She resolved to bestow on him a liberal education . . .
John Playfair, 1805

IT IS FITTING THAT JAMES HUTTON, the discoverer of the earth's
antiquity and the father of modern geology, was born in Edinburgh, the capital of Scotland. Surely no other city was more
defined by its geology. Standing not two miles from the Firth of
Forth, the finger of a bay that opens into the North Sea, the
Edinburgh of James Hutton's day was picturesque, but the
weathered topography belied a violent and chaotic antiquity. In
fact, the geologic history of the area was a textbook study of the
very processes that Hutton would later discover; it had alternated from sea to dry land at least four times in the distant past.
Geologists have now determined that volcanic activity first
occurred in the area approximately 400 million years ago. Over
the next 45 million years, the region quieted and the low-lying

plain around the volcanoes became flooded by tropical seas. A second wave of volcanism struck the area a few million years later (about 350 million years ago), forming a series of cones and raising the entire region above the sea. For the next 65 million years, the area was largely calm. However, the tropical or semi-tropical environment vacillated between low-lying forested land and flooded shallow seas. During the lengthy periods when the land was flooded, sediments formed from eroded grains and dead saltwater organisms settling at the bottom of the sea; over time, the sediments almost buried the volcanic cones. Then, at the 285-million-year mark, violence returned, again in the form of volcanic activity. This time, though, the magma did not break through the surface, instead remaining just below. The magma "intruded" into the sedimentary rocks, and was later—much later—revealed when erosion exposed it. Geologists now call an exposure of intruded magma either a dike or a sill. Thirty-five million years later (now 250 million years ago), the region was struck by a series of earthquakes, the result of mountain-building pressure that pushed and distorted the sedimentary strata. After this episode, geologic calm finally returned. Because the area was now permanently above sea level, erosion of the exposed land was the primary agent of change. Finally, 2 million years ago, the last great ice age engulfed the region and buried it under glacial ice.

Thus, all the hills that now define the Edinburgh region are remnants of the ancient volcanic activity. None is more impressive than Castle Rock. The landmark is a spectacular, nearly symmetrical cylinder of basalt; a black, shiny rock that looks almost otherworldly. Formed from magma that remained inside the ancient volcano rather than being forced out, the structure is called a volcanic plug. Today, Castle Rock rises like a broad skyscraper about 450 feet above the plain, essentially the same size at the summit as it is at the base. In fact, when one stands below and looks up, the jagged surface appears to widen as it ascends.

Castle Rock was the highest part of Hutton's Edinburgh, yet most of the old town was built on another odd geologic structure attached to the mountain of basalt. It was formed when the last ice sheet flowed slowly over the area, moving west to east, scraping across the land that came before it. It had little trouble grinding away and burying the softer rocks, but the basalt of Castle Rock was a different matter. Because of its hardness, the ice sheet was essentially forced to move over and around it. After passing around Castle Rock, the ice left a long and thin wisp of debris attached to the volcanic plug. This kind of formation is called a *crag and tail* by geologists, and it resembles a comet. When the ice age ended between 15,000 and 10,000 years ago and the glaciers melted away, the crag and tail remained as evidence of the chilly past. Today, the tail looks like a gentle uphill ramp, narrowing as it climbs, that connects the low-lying plain with Castle Rock. Near the pinnacle, the tail is a sharp ridge, perhaps only 50 yards across, the land plunging a dramatic 400 feet to the flat surface below.

As one approaches the city from the Firth of Forth to the north, or from Glasgow to the west, Edinburgh rises majestically, as if out of a fairy tale, its neighboring hills standing as sentries.

Evidence shows that Castle Rock has been inhabited continuously since Neolithic hunters and gatherers first migrated to the region 4,000 years ago. Sheer on all sides except for the ramp-like approach from the east, the mount, which has about an acre of flat rock at the top, was easily defended because the inhabitants could see for dozens of miles in every direction. They could survey the Firth of Forth to the north, and the hills to the south and east. A surprise attack was impossible. The same factors that made it attractive to the first humans made it attractive to all subsequent inhabitants of this part of Scotland, from the tribes known as the Gododdins during the Roman era to the Angles during the Dark Ages. At some point, the settlement acquired the name Edwinesburgh, probably after King Edwin of Northumbria, who ruled the region in the early seventh century. Though Castle Rock was inhabited for thousands of years, the first to build a lasting structure at the top was King Malcolm III, who built the first stone castle and chapel (still standing) after he defeated the infamous Macbeth in 1057. The town that sprang up was built completely within the castle's protective walls. After a few decades, however, the needs of the inhabitants outgrew these confines, and the first structures were built beyond the walls and on the downward-sloping ridge.

About seventy-five years after the defeat of Macbeth, construction began on the first large religious building in Edinburgh. Holyrood Abbey was situated almost exactly a mile away from the castle, at the end of the sloping "tail." Built in the 1120s and 1130s by Malcolm III's son, King David, the guesthouse of the continuously expanding religious compound was later con-

verted into the royal residence, called the Palace of Holyroodhouse. From Holyrood, the kings and queens of Scotland ruled
the country, the castle now being reserved for military purposes.
Connecting the castle and Holyrood was the town's main thor-
oughfare, High Street. The cathedral, St. Giles, completed in
1243, was built a couple of hundred yards below the castle, right
on High Street. Castle at the top, palace and abbey below, and
St. Giles in the middle of the connecting High Street—these
were the focal points of Edinburgh. By the year of James Hutton's birth, 1726, the castle was massive, its 30-foot-high fortress
walls running along the edge of Castle Rock and enclosing several large buildings that housed as many as 3,000 soldiers. Holyroodhouse consisted of several chateaulike structures and a
ruined church, the complex surrounded by well-kept gardens
and its own protective wall. High Street was paved with stone
and was very wide, especially near St. Giles and its courtyard.
The boulevard was, and still is, known as the Royal Mile.

The final defining structure of Edinburgh was built in the
early 1500s, after Scotland's army suffered tremendous losses at
the hands of the English in the disastrous battle of Flodden Field
in 1513. Fearing that the now-defenseless capital would be coun-
terattacked, Edinburgh's leaders convinced every inhabitant to
assist in building a defensive wall that would surround the city.
The wall tracked down the ridge on either side of the Royal Mile
and then closed off before Holyroodhouse. Though the Flodden
Wall was impressive—it reached heights of nearly 25 feet in
places—it ultimately did little to protect Edinburgh. A generation after the wall was built, in 1544, the English successfully

broke through and razed most of the wood-and-thatch city. However, this would be the last time large parts of the city would be destroyed. The Scots would rebuild by using local sandstone, which greatly reduced the risk of fires. Though it failed to keep out the English invaders, the Flodden Wall did serve as a tangible boundary. From now on, Edinburgh would be confined to the land within the wall.

Flodden Wall was the final defining *man-made* structure of Edinburgh. However, no description of the city would be complete without some mention of the natural forces at work. Most of Scotland is blessed with an abundance of rain, and Edinburgh is no different. The city built on Castle Rock receives rain from the North Sea and the Firth of Forth. These two bodies of water also create heavy fogs and high winds. The celebrated writer and Edinburgh native, Robert Louis Stevenson, left a marvelous description of the weather in his affectionate look at his hometown, *Edinburgh,* published in 1912:

> But Edinburgh pays cruelly for her high seat with one of the vilest climates under heaven. She is liable to be beaten upon by all the winds that blow, to be drenched with rain, to be buried in cold sea fogs out of the east, and powdered with the snow as it comes flying southward from the Highland hills. The weather is raw and boisterous in winter, shifty and ungenial in summer and a downright meteorological purgatory in the spring. . . . For all who love shelter and the blessings of the sun, who hate dark weather and perpetual tilting against

squalls, there could scarcely be found a more unhomely and harassing place of residence.

DESPITE THE INHOSPITABLE CLIMATE, Edinburgh's defensibility, and proximity to good farmland and excellent ports, inevitably made it one of Scotland's commercial centers, and therefore a city that needed to grow. However, the citizens' fear of building outside Flodden Wall ensured that expansion was held in check. Therefore, as did American builders in Manhattan centuries later, the clever builders of Edinburgh looked to the sky for the only open space. Because the city was confined to the spinelike ridge left by the glaciers, the foundations of the buildings along the Royal Mile were fragile and the tallest structures could not be more than six or seven stories high. But because the demand for space was so great, *every* building in Edinburgh was maximized and built to a height of six or seven stories. The buildings were all attached to one another, like tall row houses. Yet, only the front façades were limited to seven stories. Because the land fell so sharply away from the ridge, the rears of these buildings often had an additional three or four basement stories. The basement levels were built right into the sandstone of the ridge. Visitors to San Francisco or Pittsburgh can see similar architecture to this day.

As the population of the city grew, and as Flodden Wall kept all building activity within its interior, additional rows of buildings were constructed behind the first ones. With land so precious, the structures were built right on top of each other. Steep stairs, sloping courtyards, and narrow alleys, called *wynds,* provided the only paths to these new tenements. The old town maps from the sixteenth and seventeenth centuries look like mazes. High Street remained one of the only wide streets in the city.

The tenements were called "lands," and the way they were inhabited reflected the city's social structure. The ground floor was usually a place of business for a merchant or tradesman. Directly above the ground floor were the owner's personal quarters. Family members probably used one or two floors, depending on the size of the household. The servants and staff lived on the floors above the family quarters. And above the servants, on the top floors, would be the renters—most likely laborers such as textile workers. The lowest classes—the itinerant laborers who could find work only occasionally, the unemployed, the crippled who could not work at all, and the mentally ill—occupied the basement floors.

Though Edinburgh was always picturesque, with many citizens having views of the Firth of Forth and the surrounding hills, the geological constraints made the city severely cramped by James Hutton's time. The smells and the pollution created by the countless chimneys constantly belching smoke gave Edinburgh the disparaging moniker Auld Reekie. Moreover, the overcrowding was exacerbated by yet one other characteristic of the city that was dictated by its natural history: Because the ridge on which it was built prevented Edinburgh from having a plumbing system, an unusual system of waste removal evolved. Edward Burt, a Londoner who visited Edinburgh in 1754, left this vivid account of the "system":

When I first came into the High-Street of that City, I thought I had not seen any thing of the Kind more magnificent; the extreme Height of the Houses, which are, for the most Part, built with Stone, and well sashed; the Breadth and Length of the Street, and (it being dry Weather) a Cleanness made by the high Winds, I was extremely pleased to find every Thing look so unlike the Description of that Town, which had been given me by some of my Countrymen.

Being a Stranger, I was invited to sup at a Tavern. The Cook was too filthy an Object to be described, only another English Gentlemen whispered to me and said, he believed, if the Fellow was to be thrown against the Wall, he would stick to it. . . .

We supped very plentifully, and drank good French Claret, and were very merry till the Clock struck Ten, the Hour when every-body is at Liberty, by beat of the City Drum, to throw their filth out of the Windows. Then the Company began to light Pieces of Paper, and throw them upon the Table to smoke the Room, and, as I thought, to mix one bad Smell with another.

Being in my Retreat to pass through a long narrow Wynde or Alley, to my new Lodgings, a Guide was assigned to me, who went before me to prevent my Disgrace, crying out all the Way, with a loud Voice, Hud your Haunde (hold your hand). The opening up of a Sash, or otherwise opening a Window, made me tremble, while behind and before me, at some little Distance, fell the terrible Shower.

Well, I escaped all the Danger, and arrived, not only safe and sound, but sweet and clean, at my new Quarters, but

when I was in Bed I was forced to hide my Head between the
Sheets, for the Smell of the Filth, thrown out by the Neighbors
on the Back-side of the House, came pouring into the Room
to such a Degree, I was almost poisoned with the Stench.

In the morning, the streets were cleaned, as much as possible, by work crews that had the nastiest jobs in the city, and the
regular rains also helped to freshen up the wynds and courtyards. However, the lowest parts of the city, especially near Flodden Wall, became highly polluted.

The overcrowding and poor sanitation were certainly prob-
lematic for Edinburgh. Yet, the city's unusual layout, which
forced all residents to live close to one another, regardless of
rank and station, did have a highly positive effect. It required the
citizens of Edinburgh to be, if not egalitarian, at least tolerant.
To be sure, there was a class structure, with the rich living on
the more desirable floors of the lands and on the higher ground,
but rich and poor alike had to dodge the "terrible showers"
being thrown from the windows above. This spirit of inclusive-
ness would later inform the Scottish Enlightenment and, by
extension, Hutton's work.

THOUGH THERE IS A GREAT DEAL of information about Edinburgh
in the early eighteenth century, frustratingly few details remain
about James Hutton's early life. We do know that he was born on
June 3, 1726, to William Hutton, a merchant, and his wife, Sarah
Balfour Hutton. William passed away two years after James's
birth, leaving Sarah a thirty-one-year-old widow responsible for

raising five children on her own. James's older and only brother also died when he was young, leaving James the only male in a household that now included his mother and three sisters. We do not know the birth order of the Hutton children, but only one sister was still alive when Hutton died, so he was probably the next to youngest of the five. There is no evidence that Sarah Hutton ever remarried.

William's integrity was such that his fellow merchants elected him to the office of city treasurer, a position that he held for some time. As treasurer, William had been an active participant in the reinvigoration of a city that had suffered horribly in the aftermath of the Darien Affair, the name historians now give to the wholly private venture that nearly bankrupted the entire country. The affair began in 1698, a decade after the deposing of James II, the last Catholic king of England and Scotland, when the Edinburgh banker William Paterson convinced his fellow Edinburghers to try to colonize a part of Panama, the Isthmus of Darien. Paterson knew that the narrow isthmus was destined to be an important crossroads for world trade, and he planned to build a road across it, linked by ports on each side. This would have been Scotland's first colony. Paterson was a brilliant salesman; by tapping into his countrymen's ambitions, nationalism, and anti-English sentiment, he succeeded in persuading nearly every prominent citizen to contribute to the enterprise. Estimates vary, but it appears that Paterson raised nearly £400,000, approximately one-half of all the wealth that then existed in Scotland. Though successful in Scotland, the attempted settlement in Panama was a disaster. Spain had its own claim to the land, and her armed resistance, coupled with English obstruction and a devastating outbreak of yellow fever, caused the entire scheme to collapse tragically. Over 1,000 Scots died. Two more Scottish expeditions were sent to their doom, the brutality of the Spanish increasing each time. After two years, three expeditions, the loss of five ships, the expenditure of over £200,000, and the loss of over 2,000 Scots lives, Paterson and the Scots finally gave up. The Bank of Scotland, founded in 1695, never recovered and declared bankruptcy at the end of 1704.

The collapse of the Darien venture, and the near bankrupting of Scotland, led directly to the Act of the Union with England in 1707. Though England and Scotland had professed loyalty to one king since 1603, the two countries had remained independent, Scotland retaining its own parliament despite the English government's strong desire to unify. After the debacle of Darien, the English government agreed to pay Scotland's debts; but in return, the leaders of Scotland would have to accept the union of the two countries. The Act of the Union forced Scotland into second-class status behind England, but the nation had no choice. Its diminished standing manifested itself primarily in foreign trade—English merchants quickly dominated the Scottish textile and fish trades. In these difficult years, William Hutton was among the merchants who put Edinburgh back on its feet again, reasserting the quality of Scottish wool, linens, spirits, and paper in the Baltics and the Low Countries. By the year of James's birth, the city of 30,000 was once again thriving.

In Edinburgh, another consequence of the Act of the Union was the rise of the Whigs, a group of progressive businessmen and jurists who were eager to eliminate the excesses of antiquated feudal and mercantile laws. William Hutton and his fellow merchants were staunch Whigs: largely Presbyterian, though not usually too devout, and descended from Lowland families having ties to England. The Lowlands were the lands south of Edinburgh and Glasgow that over the centuries had been settled by invading Normans and then English. The Jacobites (followers of James II, the Catholic monarch deposed in 1688) had been the ruling class for centuries, and they were deeply suspicious of the world envisioned by the Whigs. The Jacobites embraced tradition, believed in the divine right of the monarchy to rule, and were mainly Catholic. They sprang primarily from the clans, the Scottish equivalent of feudal landowning families, who dominated life in the Highlands, the region north of Edinburgh and Glasgow that had been settled by the ancient Scottish tribes pushed north by the Normans and English. The rise of the Whigs in Edinburgh was catalyzed by the ascendancy of William and Mary to the thrones of England and Scotland (Mary was James II's daughter, but she was Protestant and not the official heir to the throne). In accepting the crowns, William and Mary agreed to greatly reduced powers, a condition insisted upon by the Whig-dominated parliaments of England and Scotland. The Whigs' consolidation of power was sealed by the Act of Union, which eliminated the Scottish parliament in Edinburgh, the last vestige of Jacobite influence.

Considering the properties that James Hutton inherited, it appears that William Hutton left his widow and children fairly well off. Certainly, the Hutton children never lacked for food or comfort. In all likelihood, William Hutton left his family in possession of land, and it was probably in the part of Edinburgh known as Lawnmarket, where the wealthiest citizens lived. It was located just below the castle at the highest part of the city, where High Street was at its widest; the view of the Firth of Forth and the surrounding hills was spectacular. St. Giles, the cathedral from which John Knox led the Presbyterian Reformation a century and a half before James Hutton's birth, was a little farther down High Street. The open courtyard in front of the church would have been teaming with activity every day as merchants concluded deals and solicitors and judges, their offices nearby, discussed the issues of the day.

After the death of William, Sarah probably rented the ground and perhaps the second floor to another merchant, or to a professional such as a solicitor. The Hutton family of five likely lived on the third floor. Most of the eighteenth-century lands are standing today, so we have some idea what the Hutton home looked like. The flat was most likely very comfortable and spacious, with enough rooms to allow for privacy for all five members of the Hutton family. The ceilings were quite high, perhaps 12 feet, and the front parlor probably had two or three windows looking onto High Street. The rear rooms may have had windows overlooking the courtyard in the back. The kitchen was large and the warmest room in the house; but there were fireplaces in nearly every room, so they would all have been well heated. Because of the active trade with the continent, Dutch tiles and cookery were common.

YOUNG JAMES WAS SENT TO THE High School of Edinburgh, located near the bottom of High Street, starting in 1736, when he was ten years old. He received the standard instruction of the day, which consisted of courses in Latin, Greek, and mathematics. Then, in November 1740, he entered the University of Edinburgh. Hutton was only fourteen, but that was the normal age to begin college at the time. David Hume, the philosopher, who was fifteen years older than Hutton, was only eleven when he started at the same university.

By 1740, the University of Edinburgh was beginning its ascent to greatness. Founded in 1583 as a divinity school during the early stages of the Presbyterian Reformation, it consisted of a series of rather decrepit buildings separated by three courtyards. Student dormitories surrounded one courtyard (though most students lived elsewhere in town). The library and administrative offices enclosed the next one, and the largest courtyard fronted the classroom buildings and faculty offices. Though in need of repair, the University of Edinburgh had an outstanding faculty, making it probably the best of Scotland's four universities (Aberdeen, Glasgow, and St. Andrews were the others).

Hutton was one of approximately 500 students at the University of Edinburgh in the 1740s. The faculty was tiny, maybe a dozen professors in all. Many fields of study that students now take for granted simply did not exist. There was no school of engineering, no economics department, no chemistry department. There was a rigid curriculum, though. The first year was devoted to Latin, the second year to Greek. Logic and metaphysics, along with natural philosophy, were the focus of the third and final year. Other than these four courses, there was but a handful of electives, which included ethics, mathematics, and history. Students paid the lecturer a fee at the beginning of each course. Though professors were paid a base salary by the university, they depended on these payments from the students to make ends meet. Thus, a professor had a strong incentive to develop a reputation as a fine speaker so that he could fill the lecture hall, and then his pockets.

It was at the University of Edinburgh that the teenage Hutton was first introduced to the ideas of Isaac Newton, which would prove enormously important to his later work. A key tenet of Hutton's theory of the earth was that it behaved like a machine, obeying constantly acting natural laws. This concept was drawn directly from Newton's natural laws of universal gravity and celestial mechanics. Hutton was first exposed to Newton in the natural philosophy course taught by Colin Maclaurin (1698–1746), one of the leading scholars on the faculty. As a young man, Maclaurin had worked with the aging Newton in London; Sir Isaac was so impressed with Maclaurin that in 1725 he wrote two letters to the university on Maclaurin's behalf, one a letter of recommendation, the other an offer to pay part of his salary if necessary. Maclaurin, a prolific writer of books and a popular teacher, viewed himself as an apostle of Newton, and he infused his natural philosophy lectures with heavy doses of Newtonian science.

Hutton was fortunate to encounter Maclaurin during his peak years, when he was particularly focused on bringing Newton's ideas to a wide audience. His greatest work, *Treatise of Fluxions* (*fluxion* was the term Newton used for calculus) was published in 1742, when Hutton was a student. This was a technical work that bolstered various propositions from the *Principia* (1687), Newton's most famous book. At the time, Maclaurin was also working on a popular book that was published in 1748, two years after his premature death. Published under the title *Sir Isaac Newton's Philosophical Discoveries,* it is still considered one of the clearest explications of Newton's ideas—ideas that would profoundly influence all the members of the Scottish Enlightenment.

Newton made at least four seminal discoveries, described collectively as the *Newtonian Revolution,* a term used even during Newton's lifetime. The first was in mathematics. Newton invented calculus (Leibnitz independently made the same discovery, and the two share joint credit for this still essential tool for scientific inquiry). He also discovered the properties of color, which led to his invention of the reflecting telescope, still in active use to this day. Sir Isaac's third great work was the mathematical synthesis of the science of mechanics, in which he defined mass, motion, inertia, and momentum. The final and most famous was his discovery of universal gravitation, to this day one of the most important scientific revelations of all time. Building on the work of Copernicus, Galileo, Brahe, Kepler, and Descartes, Newton explained how the planets, their moons, and comets maintained their orbits around the sun. Universal gravitation also explained the mystery of tides, and how objects of different weights fall at the same speed (one of Galileo's important findings).

The Principle of Universal Gravitation was the first natural law to be identified: It proved that any object with mass exerts a gravitational force, always, and that the planets maintain their orbits at all times. Another important aspect of Newton's work was his insistence on using the scientific method: building theories by accurate observation, then verifying them through rigorous yet repeatable experiments. His only nontechnical book, *Opticks* (1704), stressed the need for all scientists to follow these guidelines.

Colin Maclaurin surveyed all of the above for James Hutton and the rest of his classmates, hoping to imbue them with the same excitement that he felt for Newton's accomplishments. But Maclaurin had one other major effect on Hutton. Maclaurin was a deist, one who believes in a creator God, a God who designed and built the universe and then set His creation into motion (but does not interfere with the day-to-day workings of the system or the actions of people). The following passage is from Maclaurin's 1748 book:

> The plain argument for the existence of the Deity, obvious to all and carrying irresistible conviction with it, is from the evident contrivance and fitness of things for one another, which we meet with throughout all parts of the universe. There is no need of nice or subtle reasonings in this matter: a manifest contrivance immediately suggests a contriver. . . . No person, for example, that knows the principles of optics and the structure of the eye, can believe that it was formed without skill in that science. . . . The admirable and beautiful structure of things for final causes, exalt our idea of *the Contriver:* the unity of design shows him to be *One.* The great motions in the system, performed with the same facility as the least, suggest

his *Almighty Power,* which give motion to the earth and the celestial bodies, with equal ease to the minutest particles.

James Hutton would later use similar language in his own written works. There seems little doubt that Hutton shared Maclaurin's religious perspective. Whether Maclaurin was the only source of this belief system is not known, but it seems likely that he had a significant impact.

John Stevenson was equally influential, albeit less directly. Stevenson was a logician and he taught the third-year metaphysics course. In one of his class sessions, Stevenson alluded to the fact that it takes two acids to dissolve gold, each acid usually being powerful enough on its own to dissolve other metals. The point of the metaphor was lost on young Hutton, but the chemical process described was not. Intrigued by the idea, Hutton went in search of a book on chemistry (a course did not exist at that time), and found the only general reference work available: *Lexicon Technicum.* The chemistry described in this volume was simple, but it nevertheless attracted Hutton. From this point on, chemistry would be a singular love of Hutton's, and it would be a key tool in his later work.

Hutton graduated from the university in the spring of 1743. If he distinguished himself academically during his three years there, no record of it exists. It appears that Hutton passed his years there rather uneventfully, which is what one would expect given his young age—even the precocious David Hume graduated from the University of Edinburgh at the age of fourteen without any of his professors noticing his presence among them.

Shortly after graduation, Hutton began an apprenticeship in a solicitor's office, a job most likely secured through his mother's connections. However, sitting in a dark back room copying wills and contracts by candlelight apparently did not offer Hutton enough stimulation; John Playfair relates that "the young man's propensity to study continued, and he was often found amusing himself and his fellow apprentices with chemical experiments, when he should have been copying papers, or studying the forms of legal proceedings." Hutton's mentor quickly realized that the law was not the career for James, and he "advised him to think of some employment better suited to his turn of mind." By the autumn of 1744, Hutton was back at the university, this time enrolled as a medical student. Since medicine was the only profession of the day in which chemistry played a major role, medical school was the obvious next choice for the young man.

At this time, there is nothing to suggest that James Hutton had any interest in what would soon preoccupy him for the rest of his life: the earth. However, an unusual event occurred in the summer of 1744 that may have had some effect on him. One day, a particularly violent storm caused a landslide near the top of Arthur's Seat, which is the large mound that shadows the city near Holyrood Palace. There was no devastation or loss of life because no one lived on this formation, but the landslide created a hollow, which today is called Gutted Haddie. Exposing a large piece of the hard volcanic rock, the landslide demonstrated, in a most profound way, the power of erosion. It is almost certain that Hutton, being of curious mind, was one of the many residents of Edinburgh who went to observe the damage.

Three key features mark James Hutton's later scientific work: his application of Newtonian natural laws to the study of the earth, his innovative use of chemistry, and his recognition of the dynamics of erosion. By the summer of 1745, Hutton had been exposed to all three. Maclaurin's natural philosophy course was one of the best introductions to Newtonian science available anywhere in Europe. Hutton was now studying medicine, giving him the most intense chemistry training available. And, too, he was living in Edinburgh, whose high winds, frequent rains, and eroding mountains, monuments, and tombstones offered daily instruction in the forces of nature.

IN AUGUST 1745, THE nineteen-year-old James Hutton had completed his first year of medical school and was idling away the summer, waiting for his courses to resume in the fall. They would not begin on time. The grandson of James II, Charles Stuart, was about to lead a small army of clansmen from out of the western Highlands in a rebellion that no one could have foreseen. The uprising would do more than interrupt Hutton's education; it would deeply affect every citizen of Edinburgh and frighten the ruling Whig regime into recognizing the frailty of all they had gained for themselves and the city since the Act of Union in 1707. New forces would align to shape the future of Edinburgh, and with it, James Hutton's career.

The Storm Before the Calm

> The present unhappy troubles . . .
> *David Hume, 1745*

Except for one, they were all Macdonalds. All 150 of them. Marching in a ragged column on the dirt road headed toward the glen, this was the first time the brothers, cousins, and uncles of the sprawling Macdonald clan had seen each other fully armed. Wearing the plaids of the clan, they carried an odd assortment of old blunderbusses, pistols, swords, and pikes, most of them handed down from their fathers and grandfathers. Those without guns and swords carried sharpened farm tools such as scythes and pitchforks. Across their hips, they each wore a large bag in which they carried the provisions they would need to live in the field for two or three months.

Leading the column was the lone non-Macdonald, Charles Edward Philip Casimir Stuart. Standing a head taller than all the men, and dressed resplendently in a white coat saved for the occasion, this handsome twenty-five-year-old would soon be known to the world as Bonnie Prince Charlie. He was the reason why this small ragtag army had left their homes to fight the government; in the eyes of the Highland clans, Bonnie Prince Charlie was the crown prince of Scotland.

Charles Stuart was the grandson of James II, the last Stuart king of England and Scotland, and the son of the exiled James III. Stuart and his family had lived in Rome ever since his grandfather had been forced off the thrones of England and Scotland in 1688. Four weeks earlier, in July 1745, Charles had secretly sailed from France in a small frigate with only seven followers, and he landed in the western isles of Highland Scotland unannounced and unexpected. The prince and his confidants had made their plans clandestinely; no one, not even Charles's father, knew where he was. He had literally shown up on the doorstep of the leaders of the clans and announced that now was the time to muster the clansmen, attack Edinburgh, and regain the crown of Scotland for his father and eventually himself.

The prince had reason to believe that the Highland clans would rally around him. Twice in the past sixty years, they had formed armies to fight for the Stuarts. The first uprising occurred in 1689, soon after the Protestant- and Whig-dominated English Parliament and the Scottish Estates General had voted to oust James II and invite the Protestants William of Orange and his wife Mary to wear the crowns. The firmly Catholic Highlanders, certain that a Protestant monarchy would cause only more hardship for them, quickly rose in support of

King James. They fought several bloody battles with the English army before finally retreating to their Highland homes. The king then went into exile.

This chaotic episode was known as the first Jacobite Uprising. A generation later, in 1715, the clans again took the field for the Stuarts, this time for Charles's father, James III (known as the Old Pretender to his detractors). With France's Louis XIV supplying French troops and arms, the rebellion looked serious to the English. However, bad weather and even worse coordination conspired to doom the affair before fighting took place, and the clansmen melted back into the Highlands to wait for another day.

William Hutton and most of his fellow merchants in Edinburgh thought that 1715 marked the end of the Jacobite menace. However, Prince Charles, growing up in a household that talked of little else besides the illegal usurpation of the Stuart crown, had simply been waiting his turn. He chose the summer of 1745 to act because most of the English army was not in the British Isles; they were deeply engaged on the Continent, mired in the War of the Austrian Succession. England would react slowly at best to a disturbance in Scotland. But, unlike his father thirty years earlier, Charles was receiving no support from France. He was relying completely on the loyalty and fighting ability of the Highlanders.

This had not been the case just a year earlier. Then the French government had viewed Charles as a key player in what they hoped would be a huge British conflagration. The French intended to put Prince Charles back on English soil, backed by their own troops, to spur a Jacobite uprising. If all went according to plan, the French would place the Stuarts back on the throne of Great Britain, gaining a powerful ally while deposing a nemesis. Even if the plan ultimately failed, they would at least create a major distraction, and perhaps succeed in removing the British from French business on the Continent. It was well worth risking the lives of 7,000 French soldiers. Unfortunately for Charles, the fleet carrying the French troops was scattered by a prolonged storm, and they were never able to land. On the heels of this failure, a new set of ministers and generals gained power in France, and they had no further interest in the prince or a restored Stuart monarchy. Charles, his hopes raised so high, was now left to his own devices.

The well-worn road on which Prince Charles and the Macdonald clan were now traveling ran along the rocky and picturesque river Finnan and led to the vale of Glenfinnan. The vale was an open field bordered by tall trees at the spot where the Finnan flowed into Loch Shiel, a long, narrow lake surrounded by tall mountains. Charles had deliberately chosen this spot for the first gathering of the clans; the idyllic highland scene was meant to inspire the highland sons on whom his quest depended.

The question was this: Would there be anyone to rally? Since arriving in the western isles four weeks earlier, Stuart's original expectation of universal and enthusiastic support had been met with disappointment. The very first clan leader he summoned refused to join the rebellion, thinking it too reckless without French support. The chief of the Macdonald clan was also wary. The year 1715 was a long time ago, he told Charles, and without French backing a revolt made no sense. Still, Charles was determined, and at last he won over the chief of the Cameron clan, who said that he would be able to deliver nearly 1,000 men. With that, the Macdonalds reluctantly agreed to fight, too. The prince then sent letters to the leaders of all of the clans that he had not met with personally, and the returning messengers assured Charles that there was solid backing for his cause. So, bolstered by the allegiance of the Camerons and Macdonalds, the positive reports from the returning messengers, and the unwavering confidence of the seven companions who had sailed with him from France, Charles declared August 19 the day the fight to regain the crown would begin. And it would begin here, at Glenfinnan.

The column reached the crest of the last small hill before the woods opened to the vale, and when the prince marched through the opening, there was . . . nothing. No horses, no cannon, no bagpipes, and no soldiers. No one save his seven followers from France, who had come ahead of him, and who now stood near the lakeshore. The prince was completely bewildered. The Macdonalds entered the vale right behind him, and the chatter stopped immediately. Crestfallen, the prince walked across the field by himself to the area near the water where several huts formed a tiny hamlet. He entered one of them to deal with the shock and chagrin.

Nobody knew what to do. One of the prince's confidants went into the cottage to try to lift Charles's spirits, but he came out several minutes later looking glum. The Macdonalds scattered over the field, and it's probable that most of them thought they would be walking back to their farms the next day. Finally, after an interminable wait of two hours, the group heard the sound of bagpipes in the distance. The music drew closer, and with their eyes trained on the hill near the entrance to the vale, the Macdonalds saw first the heads, and then the bodies, of the huge Cameron clan. They were marching in two columns, and their order was impressive. They reached the field and kept marching toward the lake. The Macdonalds quickly got back into ranks, not wanting to be outdone. When Charles heard the bagpipes, he came out of his refuge, and watched excitedly as the columns advanced. He assembled his seven followers and moved to the small knoll adjacent to the lake, whereupon the 800 Camerons and the 150 Macdonalds formed a semicircle in front of the prince. There was nothing to be embarrassed about now.

Still, a prudent man would have taken measure of the situation and returned to France. Instead, and inexplicably, with only 950 men before him, Charles instructed his highest-ranking companion, a gentleman who called himself the Marquis of Tuillibardine, to unfurl the standard of the Stuarts, which was a red, white, and blue silk flag, and read the declaration of war against "the Elector of Hanover," a taunting reference to George II, whose German House of Hanover now ruled England and Scotland. The rebellion had begun.

CHARLES STUART WAS CHARISMATIC, but he knew nothing about leading an army. In the hours after the reading of the declaration of war against King George, another 450 or so Highlanders made their way to Glenfinnan before nightfall, creating a force totaling 1,400 men. George Murray, an experienced officer who had seen a great deal of action in Europe as a mercenary, was among them. Murray was quickly designated as the military commander of the Highlanders. He and the other clan leaders decided to march straight to Edinburgh. By moving quickly toward their ultimate goal, they hoped to surprise the government forces. Other clansmen would no doubt join them during their march.

Meanwhile, in Edinburgh, rumors began circulating in early August that Charles had landed in western Scotland. No one believed them at first. But reliable intelligence finally arrived from the west. General Jonathan Cope, the ranking military officer in Edinburgh, was charged with outfitting an army and eliminating the nuisance immediately. Though Cope had been in the king's service for over two decades, he had seen little military action. One commentator later pointed out that, "From this point onwards, [Cope's] incapacity for high command showed itself flagrantly." Cope, like everyone else in Edinburgh, was confident that Charles could not possibly raise a large body of clansman, so the alarm was muted.

Not taking any chances, though, Cope decided to engage the Highlanders as quickly as possible, before their ranks were able to swell. Yet he was able to muster only 1,400 troops in Edinburgh because most British soldiers were on the European continent. Reluctant to wait for more men to arrive in the city, and expecting that his forces would increase as they marched through pro-government territories, Cope and the government's army departed Edinburgh on August 22. Without either side knowing the situation of the other, the two small but equal-sized legions began marching toward their first engagement on practically the same day.

That first encounter occurred about 50 miles from Edinburgh, at a mountain known as Corriearrick. Cope was preparing to engage the enemy there, when at the last moment he received faulty intelligence indicating that the Highland infantry was much bigger than his. He disengaged and led his troops north to Inverness, 160 miles from Edinburgh. This meant that the way was clear for Charles and the Highlanders to march into the former capital.

When news of Cope's shocking nonengagement reached Edinburgh, the citizens realized that they had been left unprotected. A hastily called town meeting made it clear to all just how unprepared the city was. There were no troops to prevent the highlanders from capturing the city, beyond the 600 elderly "soldiers" that Cope had left behind to guard the castle; the officer in charge was eighty-five years old. At the meeting, the majority voted to do nothing, praying that the Highlanders had no interest in destroying Edinburgh and fearful that their sons, if hastily assembled into a fighting force, would stand no chance against the ferocious clansmen. In addition, some townsfolk exhibited a lackadaisical attitude that revealed their pro-Jacobite leanings—clearly a larger number of Jacobites lived in Edinburgh than the Whig majority realized.

However, two town leaders would not accept such acquiescence, and they announced that they would take it upon themselves to raise volunteers and defend the city. One of the leaders was none other than Colin Maclaurin, Hutton's professor of natural philosophy. The other was George Drummond (1687–1766), a past mayor (called the Lord Provost), who had founded the University of Edinburgh's medical school in 1726, the year of Hutton's birth. Together, they enlisted 400 volunteers to defend the city, mainly students from the university. It is unclear whether James Hutton was among them. He and his mother and sisters could have been at the family farm 40 miles southeast of Edinburgh. In fact, many citizens of Edinburgh who had family in the country left the city when word of the Highlander threat first arrived.

On September 15, scouts in the countryside delivered the news to the leaders of Edinburgh that Prince Charles and the Highlanders were only 8 miles away, half a day's march. Maclaurin and Drummond mustered the volunteers in the Lawnmarket area, near the castle, and then Drummond led them down High Street to take up positions at the town gate at the bottom of the ridge. Inexplicably, the crowd along the street, who were there, the volunteers surely assumed, to cheer them on, instead peppered them with jeers and catcalls. In the face of this antagonism, the fragile resolve of the young men disappeared. One by one, the volunteers quietly left the ranks and melted into the crowd. By the time Drummond reached the town gate, there were perhaps a dozen boys left. His army had become a brigade. This is when he gave up—furious with his fellow citizens—and let the last of the students go home.

The next day, an advance troop of Highlanders entered the city through the town gate unopposed. They quickly marched up High Street and secured the rest of the town. Not a shot was fired, and there were no casualties. The 600-strong garrison locked itself in the castle, and remained there for the rest of the occupation.

Then, on September 17, Prince Charles Edward Stuart, son of James III and grandson of the last Stuart king to reign in Great Britain, entered the city and took up residence at the Palace of Holyroodhouse, the very spot from which his predecessors had ruled Scotland. Without losing even one soldier, and in just eight weeks from his arrival in the Highlands accompanied by only seven friends, the prince had taken over the capital of Scotland.

Charles, Murray, and the clan chiefs knew that Cope and his army would appear soon. After retreating to Inverness, Cope had marched his 1,400 men east to the port of Aberdeen. There they boarded several naval ships and sailed down the coast and across the mouth of the Firth of Forth to the town of Dunbar, which was about 30 miles east of Edinburgh. Cope was now marching toward the capital. But, in occupying Edinburgh, Charles had a major advantage: He could choose where the inevitable battle would be fought. George Murray did not want to face the government soldiers in the city itself, so he chose a field near the town of Preston, just to the east of the city, as the spot for the encounter. Cope's army would be marching through Preston on the way to Edinburgh. The night before the expected confrontation, the clansmen vacated Edinburgh to take up their positions.

The accepted method of battle in the mid-eighteenth century was for the two infantries to face each other on an open field, get off one round of musket fire (neither side had cannons), and then charge each other with swords and pikes. This practice had barely changed for hundreds of years. The Highlanders, especially those from the Cameron clan, used a particularly nasty weapon for want of guns: a sharpened scythe, which was a long pole with a curved blade attached. Farmers used this tool for harvesting, but in battle it was an opponent's nightmare. If a Highlander got off a solid stroke of the scythe at close range, he could quite literally cut an enemy in half.

The battle of Preston lasted only thirty minutes. The soldiers awoke on the morning of the expected engagement to find themselves in a thick fog, a common occurrence in this area so near the Firth. The fog allowed the Highlanders to begin the attack before Cope's troops were completely ready. Eyewitnesses of the battle, and there were many, describe a gruesome encounter. The clansmen were able to reach government lines quickly, wielding their heavy swords and scythes ferociously. They sent Cope's men into flight, but not before there was awful bloodshed on the battlefield. One eyewitness describes an almost unimaginable scene, the field strewn with sliced-off legs, arms, hands, and even heads and torsos. Those who lost an appendage lay bleeding to death, probably in silence due to shock. Others were screaming, trying to keep their entrails from spilling out onto the field. There was so much blood on the ground that it appeared as if a red rain had fallen. One government soldier had raised his arm to block a Highlander's razor-sharp broadsword, which then cut off his hand before slicing halfway through his skull.

Even Charles was appalled by what he saw; in a letter to his father, he described the horror of watching so many young men, all his countrymen, horribly butchered. Approximately 500 government soldiers were killed, and most of the rest were taken prisoner. Cope's army no longer existed. The Highlanders, by comparison, had suffered fewer than three dozen casualties.

With the victory at Preston, the prince was now the undisputed leader of Scotland. Now what? Realistically, no one had expected Charles to get this far. But here he was in possession of Edinburgh, a solid army was still intact, and he had plenty of food, money, and able advisors. Murray and the leaders of the clans advised the prince to be content with the capture of Scotland and to prepare to defend Edinburgh and Scotland from the English forces that would eventually come from the south. But the prince believed that England was also part of his legacy, and he insisted that his men march into England, too. Murray and the clan chiefs reluctantly agreed, acknowledging that perhaps the best defense against counterattack was to move aggressively against their southern neighbor.

They would not do so immediately. It made sense to wait and allow more Highlanders to join the force already in Edinburgh. They also wanted to see whether the mercurial French government would change its mind and send support. In just a few weeks, the Highland army swelled to about 5,000 men. In addition to soldiers, Jacobite supporters who were not expected to fight—older men, women, and children—came from the surrounding countryside to be part of the new Stuart regime. Reports of the period are fascinating. Charles held court and behaved like a monarch, holding at least one royal ball and even announcing several edicts. His demeanor, though, was said to be modest and controlled. The most humorous accounts focus on the behavior of the young women in the city, primarily the daughters of Jacobite supporters. They apparently fawned over the young and handsome prince.

The respite did not last. Unable to wait any longer to hear from the French, in early November the army of Bonnie Prince Charlie was once more on the move, heading south into England. They would need to reach London before the heavy rains of winter commenced. Just as on their march from Glenfinnan to Edinburgh, the Highlanders met almost no resistance for hundreds of miles. On November 9, they took the town of Carlisle, just across the Scots/English border. By early December, the army had reached Derby, over 200 miles south of Edinburgh, and only 130 miles away from London. The Highlanders appeared unstoppable.

However, Charles's ultimate success depended on two factors. First, he needed the Catholics and Jacobites of England to join his standard and swell the ranks of his army of 5,000. Instead, they offered no support at all, either out of indifference or certainty that the prince's effort was folly. Second, he desperately needed the help of the recalcitrant French. Impressed by what Charles had accomplished, the French at this point were, in fact, hurriedly trying to form a small expeditionary force, but it would be several weeks before it would arrive, and then it would be too little too late. On December 4, 1745, Charles and Murray met with the rest of the clan leaders to decide their next step. They knew from firm intelligence that an English force of around 30,000 men, six times their size, was on the march from London. The Highlander chiefs voted to retreat back to Scotland. Charles argued vehemently to continue to London, but he bitterly caved in when Murray, whose leadership had been so impressive, voted with the rest of the Highlanders. The beginning of the retreat was the beginning of the end, not only for the Stuart cause, but for old Scotland as well.

Murray, who kept the ranks together and moving quickly, handled the retreat masterfully. They crossed back into Scotland in early January. Several days later, the Highlanders fought their second battle against government forces, this time at Falkirk. In a driving winter rainstorm, the clansmen once again chased the better-trained government troops from the field. The Highlanders could not take much from the victory, however. They could not go back to Edinburgh because a new government army, formed in Glasgow, had retaken and secured the city just a few days before.

The English army was getting reinforcements from the south daily. The Highlander army, on the other hand, was losing men to desertion at a steady rate. The clansmen had been in the field for six months now, and the winter weather, coupled with the realization that England had now made the Highlanders' defeat a high priority, made it harder and harder to keep all but the most committed fighting. Therefore, the retreat continued to the historic capital of Highland Scotland: Inverness. Bonnie Prince Charlie and his men remained there for the rest of the winter, preparing for the showdown with the government forces they knew would come with the beginning of spring.

The engagement occurred in early April 1746. Murray and Charles made the decision to meet the enemy at Culloden Moor, just outside Inverness. The battle pitted the Highland force of about 4,000 against a foe at least double that size.

More crucially, the government army now possessed numerous cannons. The battle began in the early afternoon, and though the Highlanders fought valiantly, it was over quickly. In less than an hour, over 1,200 clansmen lay dead or dying on Culloden Moor. The order was given to retreat. Though the losses were severe, Charlie's army was still basically intact.

What happened next, though, was unexpected, and completely foreign to the rules of war. To this day, it is remembered by all Scots. When it was clear that Murray's men were in retreat, the commander of the government forces, the duke of Cumberland (who was the younger brother of King George II, and therefore had a personal stake in the destruction of the Jacobites), ordered his soldiers to spare no one, not even the wounded lying in the fields and woods. "No quarter" was the order given. Hundreds of the fallen were shot or stabbed where they lay. Some were even buried alive. Many of the captured were shot on the spot. Those who were not killed were thrown into prisons.

After the butchering of the Highlander army at Culloden, the atrocity continued. Cumberland, with the complicity of the government in London, had decided that a repeat of this rebellion must never occur. He was going to "pacify" the Highlands once and for all. Several days after the battle of Culloden, the government army scattered throughout the Highlands, bringing destruction in their wake. Cattle and sheep were slaughtered, crops ravaged. Cottages, farms, and houses were burned in every district of the Highlands. The lands of the fallen clan chiefs were forfeited and turned over to special managers from the Lowlands. Laws were quickly passed that stripped the chiefs of all authority. Clan councils were declared illegal, as was the wearing of tartans, the playing of pipes, even the mere speaking of Gaelic. The "harrying of the glens," as the pacification came to be known, was thorough, cruel, and brutal. Butcher Cumberland was singularly successful in ensuring that the clans would never rise again.

BONNIE PRINCE CHARLIE ESCAPED to the western islands of Scotland and then to the Continent, where he lived out the rest of his long and besotted life (he died in 1788). For the bonnie prince, "the 45" was a bitter disappointment, yet he was able to resume his privileged life in exile upon his return to Italy. Would that the clans could have returned to their former lives. The vicious reprisals against them forever changed the landscape of Scotland. The clans never recovered, and Highland culture became just a memory. Even Lowland Scots realized that something was lost with the passing of the clans. But what was lost in their eyes did not outweigh what was gained—the elimination of the fear and wariness caused by the Highlanders and their violent and martial culture, their arbitrary and antiquated laws, and the genuine risk of real conflict. The 45 was the third major armed encounter in as many generations. A modern society, which the Whigs were trying to create, could not achieve its full potential with this sword of Damocles hanging over its head. So, although Edinburgh's Whigs did not approve of the way the Highlanders were quelled by the English government and army, they were pleased with the result—the end of the specter of violence. An energy and determination fell over the city that laid the groundwork for an extraordinary intellectual flourishing a generation later, one of the leading participants of which would be the now somewhat aimless student, James Hutton.

Youthful Wanderings

> But, happily, the force of genius cannot always be controlled
> by the plans of a narrow and shortsighted prudence.
>
> *John Playfair, 1805*

W HEN BONNIE PRINCE CHARLIE and his troops left Edinburgh on November 3, 1745, to begin their march on England, they left only a small force of 500 men to maintain Jacobite control. Two months later, while Charles's forces were bogged down near Stirling Castle, English soldiers retook Edinburgh, forcing the remaining Highlanders to flee without firing a shot. One of the first official acts of the royal army was to find Archibald Stewart, the Lord Provost, and throw him into prison. Stewart was accused of aiding and abetting the enemy because he had presided over the meeting in which the town leaders had voted to do nothing to stop the rebels. George Drummond, the fifty-nine-year-old former Lord Provost and one of the two city leaders who had mobilized to defend Edinburgh against the clansmen, was installed in his place. The Whig-Jacobite tension that had simmered below the surface prior to "the 45" could no longer be ignored. The defeated Jacobites, like Stewart, were now stripped of all influence.

The confusion and uncertainty that existed in Edinburgh in 1746 were mirrored in James Hutton's life. For the next eight years, he would bounce from one enterprise and location to the next, but fortunately all the while adding to his storehouse of scientific knowledge.

IN THE WINTER OF 1746, WITH THE Highlanders no longer walking the streets and drinking in the taverns, the citizens of Edinburgh tried to resume some semblance of normalcy. The battle of Culloden would not occur for a few more months, so there was still tremendous unease in the city and throughout Scotland. James Hutton would do his part to get on with life by resuming his medical studies.

The university was founded in the late 1500s and was among the oldest in Great Britain, but the medical school was quite new; it was founded by George Drummond the year Hutton was born. Drummond had high ambitions for the town, and he viewed the University of Edinburgh as an institution that could have a positive influence on other parts of the city. At the time, there was no medical school in Scotland, so any young Scot wanting to become a doctor had to study on the Continent, or else declare himself an Anglican and try to gain admittance to Oxford or Cambridge. This situation was absurd in Drummond's eyes, so he pressured the administrators of the university to hire Alexander Monro as the first professor of anatomy; Monro, in turn, hired four additional faculty members. Drummond next forced the city council to approve the building of Edinburgh's first infirmary. His vision was for the two new institutions to be integrated, and the medical school quickly became known for requiring its students to work bedside with sick patients immediately after they began their instruction.

If Drummond saw the need for the integration of theory and practice, in Alexander Monro he had the ideal scholar to make this vision a reality. Monro and the other four original faculty members had studied medicine under Hermann Boerhaave (1668–1738) at the University of Leyden, in Holland. Boerhaave is recognized as the first modern medical doctor in Western history, for he stressed the need to observe directly, to diagnose, and then to treat patients in an era when "doctors" usually kept a safe distance from their patients. A gifted writer and lecturer, Boerhaave attracted many to Leyden and made it the center of medical training in Europe. The curriculum he created wove together strands of anatomy, physiology, chemistry, and pathology. Postmortems were made routine, which was a major innovation. As a follower of Newton's, Boerhaave also looked for general laws about the systematic operation of the body and the progress of disease within it. Monro brought this sensibility with him to Edinburgh, and the new medical school was modeled on Leyden's. Thus, any student enrolled in Edinburgh's medical school when Hutton was there would have received intense instruction in anatomy, chemistry, and

Newtonian science, along with an appreciation for the importance of direct observation over slavish adherence to theory.

The medical students were practically buzzing with anticipation at the end of 1747. The popular professor of physiology, John Rutherford, was about to begin clinical lectures in the infirmary's operating room. This development represented a significant departure from past practices and was eagerly awaited by the students. However, instead of participating, Hutton left for Paris, where he continued his medical studies at the 700-year-old University of Paris. John Playfair rationalized Hutton's action by pointing out that the medical school in Edinburgh was still not established as a world-class institution, so it was common for students to finish their studies on the Continent. Still, it is more likely that Hutton left Edinburgh not because the medical school was deficient but because he was advised against staying. Sometime in 1747, it seems, Hutton impregnated a young woman. No one, except perhaps for Hutton's family, knew about the illegitimate child until after Hutton's death.

In E. C. Mossner's standard biography of David Hume, there is a marvelous passage about how illegitimacy was dealt with in the eighteenth century. Apparently, David Hume's own father, Joseph, impregnated a young servant girl when he was twenty-one, the same age as Hutton when his indiscretion occurred. The young woman, Elspeth Burnett, who was a servant to Joseph's uncle, testified before a church committee that she was with child in March 1702, and she claimed that Joseph Hume was the father. Joseph took his time about appearing before the same committee. When he finally showed up in

August (the baby probably had been born by then), he simply denied the accusation. With barely a pause, he announced that he had to leave for Utrecht, in Holland, and off he went. No one tried to stop him. He did, in fact, go to the University of Utrecht to study law, and he stayed in Holland for the next three years. Mossner points out that young men in Joseph Hume's position were usually recommended by family members to leave town and stay away for a while until the "affair" was forgotten. It is likely that the Hume family paid the girl a modest amount of money and made a contribution to the church poor box. Then the matter was dropped, and "the scandal would probably not have damaged his own good name irreparably." By leaving Edinburgh for Paris, Hutton appears to have been doing what any young man in his situation would have done.

Hutton remained in Paris, where he "pursued with great ardour the studies of chemistry and anatomy," according to Playfair, for a little over a year. The University of Paris was among the oldest in Europe. The date of its founding is obscure, but it was a formal institution of higher learning from the eleventh century on, and its medical school was probably the oldest in Europe. Paris would have been incredibly exciting, both culturally and aesthetically, for the young Hutton. The gardens, the cafes, the wide boulevards were already a feature of the city, thanks to the long and prosperous reign of Louis XIV (1643–1715), and the open spaces would have been a welcome departure from Edinburgh's narrow wynds and generally overcrowded conditions. The unrest that would lead to the French Revolution two generations later was not yet openly manifesting itself. Rather, this was the Paris of

Montesquieu, Diderot, Voltaire, and the young Rousseau—the height of the French Enlightenment.

Given Hutton's pronounced interest in chemistry, it is quite possible that he attended several of the riveting lectures given throughout the year by Guillaume-François Rouelle (1703–1770), a chemist at the Jardin du Roi. Scholars at the Jardin were obligated to teach public courses, which were advertised and well attended. Rouelle was reputed to be a gifted public speaker with a showman's style, particularly when it came to demonstrating chemical experiments. Antoine Lavoisier, the father of modern chemistry, was said to have been inspired by him. Rouelle had a fascination with the chemical makeup and structure of the earth's surface, and gave at least one lecture on the subject.

After his year in Paris, James Hutton packed up once again and moved to Leyden, in Holland, where many Scots finished their medical schooling. Not even ten years since Boerhaave's death, the medical school was still the finest in Europe. Leyden, though much smaller than Paris, was a vibrant city with a thriving textile trade. Leyden was only thirty miles from Amsterdam, where Dutch culture was enjoying its zenith.

In September 1749, after five years of study, James Hutton finally received his medical degree. His University of Leyden thesis was titled *De sanguine et circulatione in microcosmo (The Blood and the Circulation of the Microcosm)*. This thesis was significant because it made use of Newton's notion of cycles (as seen in the orbits of the planets) in analyzing the circulatory system, which is what allows the human body to be self-sustaining. Hutton would later think about the earth in much the same way.

IN THE FALL OF 1749, SIX YEARS after having failed as a solicitor's apprentice, James Hutton was finally ready to begin a career. He had a degree from the best medical school in Europe and had studied at two other outstanding institutions. With nearly a decade of schooling behind him, he had received the finest available instruction in chemistry, by way of the study of medicine, and had been imbued with Newtonian thinking, thanks to Colin Maclaurin. His medical training had also honed his observation skills. But instead of going back to Edinburgh to begin a medical practice, he moved to London. Perhaps he wanted to avoid Edinburgh because of his illegitimate son. Or perhaps the opposite was true—he was helping the mother of his son establish herself in London (the boy was known to have spent most of his life there). Whatever the reason, we know that from his residence in London, Hutton wrote to his friends in Edinburgh and expressed concern that it would be difficult to start a medical practice back home. Yet he did not try to begin a practice in London, either. In fact, after all the years of preparation, there is no evidence that he ever seriously considered practicing medicine.

One of the acquaintances to whom Hutton wrote was a former classmate named James Davie. Hutton and Davie had worked on chemistry experiments together during their school years. As students, they had found a way to make the chemical sal ammoniac, which was used as a flux in metalworking (to connect two pieces of metal). Davie and Hutton had discovered how to make the flux from common coal soot, an abundant substance in any northern city of the age. Thus, the expense for raw materials was essentially nothing—one merely paid chimney sweeps a few trifles for soot. Moreover, at the time, the only available sal ammoniac came from Egypt (where it was made from camel dung). Davie wrote to Hutton suggesting that perhaps they should try to sell sal ammoniac by using their method. Hutton left London for Edinburgh a few months later, in the summer of 1750, and worked with Davie to set up the chemical works. No details of the business arrangement exist, but it appears that Davie made Hutton a partial partner at this time; Hutton was made a full partner in 1765. The men loyally adhered to the arrangement for the rest of their days. The firm was an immediate success, and it provided Hutton with a steady income; this, combined with his inheritance, made him financially comfortable.

The sal ammoniac work is the first tangible evidence that Hutton was an unusually gifted and original chemist. It was his knowledge of chemistry that separated him from most of the other early geologists and allowed him to produce such an original theory. Many scientists understood Newton's teachings, and many also were keen observers of the natural world, but few early students of the earth had the gift of chemistry. In the next stage of his life, Hutton would continue to pursue his chemistry experiments while starting to pay attention to geological processes. It is possible that the success he and Davie had in isolating an important chemical from a mineral source—soot being a by-product of coal—helped to combine these two passions.

Not much is known about the specifics of Hutton's life in the early 1750s, but clearly these were critical years. First, he made the decision not to pursue a career in medicine. Then, James Davie appeared seemingly out of thin air to provide him with financial security. And finally, after several visits to his farm (which he had inherited from his father) some 40 miles southeast of Edinburgh, he made the decision to devote himself to farming. Given that Hutton had spent his life in cities, this was an enormous change, but it was key for his future scientific work. He would start thinking deeply about the land he farmed, which would help prove his theory about the ancient earth over three decades later.

IF HUTTON WAS GOING TO BE A FARMER, he wanted to be a modern, innovative one. But he quickly learned how difficult that would be. At this time, agricultural practices in Scotland were very backward, having changed little over hundreds of years. In 1752, probably on the advice of John Hall, an influential friend who lived near his farm and who was the future father of James Hall, James Hutton made yet one more move. Certain parts of England were known for their agricultural innovations, and one of those areas was Norfolk, a county north and east of London. Hutton persuaded a farmer named John Dybold to let him live and work on his farm for a short time. That "short time" stretched to two years.

These were happy and instructive years for Hutton. As John Playfair later wrote:

He appears, indeed, to have enjoyed this situation very much: the simple and plain character of the society with which he mingled, suited well with his own, and the peasants of Norfolk would find nothing in the stranger to set them at a distance from him, or to make them treat him with reserve. It was always true of Dr. Hutton, that to an ordinary man he appeared to be an ordinary man, possessing a little more spirit and liveliness, perhaps, than is usual to meet with. These circumstances made his residence in Norfolk greatly to his mind, and there was accordingly no period of his life to which he more frequently alluded, in conversation with his friends; often describing, with singular vivacity, the rural sports and little adventures, which, in the intervals of labour, formed the amusement of their society.

During his tenure in Norfolk, James Hutton started thinking about the earth in a rigorous way. He traveled extensively around England, and later the Continent, observing farming methods *and* geology. In a letter to John Hall, Hutton revealed that during his hikes he found himself examining the surface of the earth, and looking in every pit, ditch, or bed of river that fell in his path. He would continue making field excursions for the rest of his life; he could later claim to have seen practically every corner of England and Scotland.

THE STUDY OF THE EARTH WAS IN its infancy in the mid-1700s—the term *geology* would not come into use for another generation. Indeed, the first chapter of the Book of Genesis stated precisely how the earth had been created, so for most Christians there was no need to inquire further. Yet, for the scientific community, alive with activity since Galileo's and Descartes's work of the early seventeenth century, numerous questions about the earth needed to be answered. For example, how did a volcano work, what caused earthquakes, and what were those objects that looked like organisms mineralized into stone (that we today call fossils)? Notably, though, all early studies of the earth attempted to be scientifically rigorous while still deferring to the time scale dictated by the Bible, and stressed the central role of Noah's Flood and the waters of the newly created earth.

If Hutton had told one of his old professors at the University of Edinburgh of his newfound interest in the earth, and then asked him for a list of the key books in the field, the professor would likely have recommended nine works, all of which were popular or influential at one time. They fell into two groups, the first being investigations of specific earth processes (e.g., earthquakes), the other being all-encompassing "cosmogonies" that attempted to understand *everything*—how the earth was created, how it evolved, how it was going to end, and so forth. This small library encapsulated geologic thought, circa 1752.

Hutton would have been directed to begin with Nicolaus Steno's (1631–1687) *Dissertation Concerning a Solid Body Enclosed by the Process of Nature Within a Solid,* published in Italy in 1669. This book made two important strides, and is now widely regarded as the first rigorous work in modern geology. First, it properly identified a fossil as a once-living organism that had died on freshly deposited sediments, was buried by more sediments, and was then petrified (before Steno's pronouncement, fossils were thought to be "figured stones"—minerals that through pure happenstance formed in shapes that resembled organisms). Second, the work carefully and correctly described how sedimentary rocks formed. Steno realized that all sediments accumulate in horizontal layers at the bottoms of bodies of water and that the bottom layer of a sedimentary rock formation must have formed before any of the layers on top of it. Like Hutton later, Steno realized that the rock record he studied did not match well with the standard interpretation of the Bible, yet he was not prepared to jettison the Scriptures, as Hutton was. Instead, he devised a scheme of six stages that explained his observations while still keeping the earth under 6,000 years old. He believed that after Noah's Flood, a second cataclysm must have occurred, in Italy at least, to account for the formations he saw there. Hutton probably read the 1671 English translation of Steno by William Oldham, a distinguished member of the Royal Society who viewed the work as seminal.

The next crucial book for a newly inspired student of the earth would have been Robert Hooke's posthumously published *Discourse of Earthquakes* (1705). Hooke (1635–1703) was an extraordinarily talented and influential scientist, second only to Newton in late seventeenth-century England. He concurred with Steno's description of how fossils formed, and suggested that violent upheavals, such as earthquakes, had raised undersea sediments above sea level in the past, which was correct. However, as it was for Steno, Hook's earth was only as old as the Bible allowed.

After Steno on fossils and sediments, and Hooke on earthquakes, Hutton would have been urged to read Anton-Lazzaro

Moro (1687–1740) on volcanoes. His *De' crostacei* was published in Venice in 1740. Though it demonstrated an impressive understanding of the power of volcanic action and observed that lava came from deep within the earth, the overall theory was still strongly tied to the Book of Genesis.

The other six books were all cosmogonies, their authors attempting complete histories of the earth (Hutton would later strongly object to this type of effort). The first three scholars, all from England, started a tradition that was later called "biblical geology," that is, the effort to link scientific laws to biblical history. The most ambitious book was Thomas Burnet's *The Sacred Theory of the Earth*, published in four volumes from 1681 to 1689. As the title implies, Burnet's work treated the Holy Scriptures as the starting point. It was a long, very complicated book that sought to explain the evolution of earth in the context of Newton's laws while adhering strictly to the Bible. In astounding detail, Burnet explained how the earth started out as a paradise with a mild climate everywhere, only to be distorted by the great Flood. He continued by projecting what the future held for the earth, and predicted that a planet-engulfing fire would send us all to a fiery death. Burnet was part of the English scientific establishment, Edmund Halley and Isaac Newton being close friends. Just as Newton's chronology helped to keep biblical chronology viable for another century, his assistance and endorsement of Burnet's book helped to keep the study of the earth wedded to that chronology. Newton and Burnet had a lengthy correspondence over two trouble spots in the book. Burnet wanted to, and ultimately did, start his book with the

Flood, and then go back to the newly created earth. Newton, on the other hand, believed that the starting point should be Creation itself because he thought Noah's Flood only further distorted what had already existed. And Newton thought that Burnet's trouble with just six twenty-four-hour periods for Creation could be solved by supposing that the earth had rotated more slowly in the past (therefore a day was longer than twenty-four hours). Remarkably, Newton was prepared to compromise on the natural laws that he had discovered.

John Woodward's *Essay Toward a Natural History of the Earth* was published in 1695. Inspired by Steno's recognition of the true nature of fossils, Woodward was particularly eager to explain the remarkably wide distribution of them around the world. He argued that the Deluge destroyed the original surface of the earth and, in the process, caused the scattering of living creatures that were later found preserved as fossils.

William Whiston's *New Theory of the Earth* (1696) was written in the same spirit as Burnet's, but he worked harder to come up with scientific explanations for known phenomena. For example, he pronounced that the Deluge was caused by a collision of the earth with a comet. Invoking comet collisions was popular at the time because a remarkable comet had been visible in England and Europe for most of 1680.

After the appearance of these important works at the end of the seventeenth century and the beginning of the eighteenth, there was a quiet spell that was broken right before Hutton's move to his farm. Within a couple of years of each other, three distinguished works were published. The first was the posthumous publication of Leibnitz's (1646–1716) *Protoggea,* published in 1749. Taking a cue from René Descartes, Leibnitz was the first to propose an idea that would remain popular for the rest of the century and would cause great frustration for James Hutton: a universal ocean. The universal ocean was a different concept from Noah's Flood, yet it was a variation on the same theme. Leibnitz argued that soon after the earth formed, it was covered by a great sea that eventually dissipated to reveal the already-complex continents.

Another posthumously published book, Benoit de Maillet's (1656–1738) *Telliamed* (the author's last name spelled backward, 1748), was actually the first to propose that the earth was ancient—about 2 billion years old in the author's estimation. Maillet also envisioned an earth once completely submerged by a universal ocean, and he calculated the 2 billion years based on his analysis of how fast the waters were receding into vortices. The book went through three editions in French and at least one in English. Though a popular read, in the end Maillet's science was unpersuasive. This was because the book was supposedly written by an Oriental philosopher and based on Egyptian legend (Maillet probably chose this format to avoid the censure of the church), and it contained several wild claims; for example, it proposed that all species of organisms began as seeds (found throughout the universe), and that women and men had been transformed from mermaids and mermen.

However, Maillet's work did inspire the last important book of the period, G. L. de Buffon's thirty-four-volume *Histoire Naturelle* (1749). Buffon was the Intendant of the Jardin du Roi in Paris, an influential position. In his widely read volume about the earth, Buffon argued that it and all the planets had formed after a collision between the sun and a comet or comets. The earth therefore started out as extraordinarily hot. Over time, an ocean formed that engulfed the entire planet (another universal ocean), and the recession of the waters led to the composition of the land now inhabited.

In January 1751, a couple of years after the publication of his ambitious work, Buffon received a letter from the faculty of the Sorbonne. The missive informed him that they had found fourteen ideas in his volumes that were "reprehensible and contrary to the creed of the church." The key offending lines were these: "The waters of the sea have produced the mountains and valleys of the land—the water of the heaven, reducing all to a level, will at last deliver the whole land over to the sea, and the sea, successively prevailing over the land, will leave dry new continents like those which we inhabit." Apparently, the faculty disapproved of Buffon's implication that God did not create the mountains and valleys directly; rather they were the result of secondary causes—the waters of the sea.

To keep his prestigious position, Buffon had to recant: "I declare that I had no intention to contradict the text of Scripture; that I believe most firmly all therein related about the creation, both as to order of time and matter of fact; and I abandon every thing in my book respecting the formation of the earth, and in general all which may be contrary to the narration of Moses."

Clearly, as Hutton began his earnest inquiries into the earth in 1752, the field was still deeply influenced by the Book of Genesis. The extraordinary hold of the Bible prevented genuine freethinking about the history and workings of the planet, and the few open-minded scientists who did emerge were quickly censured by the church. Though Steno, Leibnitz, and Buffon were skeptical about the earth being only 6,000 years old, they did not openly confront the issue.[1] Only Maillet published an argument for an ancient earth, but because the author was already dead, and the flaws of his book were so pronounced, the claim had almost no power. There was really only one truly seminal and lasting work from which Hutton could build his own theory, and that was Steno's. He would certainly use it as a starting point. In time, he would build a remarkable edifice on that foundation.

---

[1]Buffon would later revise his book and state that the earth was 75,000 years old. He arrived at this number through experiments that he conducted to estimate the cooling rate of the earth. The revision was less influential than the original, though, because when it was published in 1778, geologists were already under the influence of Abraham Werner's theory. Also, several of the assumptions of Buffon's hypothesis were no longer believed valid thirty years later.

The Paradox of the Soil

At last he fixed on his own farm in Berwickshire, and accordingly
set about bringing it into order with great vigour and effect.
John Playfair, 1805

IN THE SUMMER OF 1754, AFTER spending two years learning
about farming and husbandry with John Dybold in Norfolk, the
peripatetic James Hutton began a period that could not have
been more settled. For the next thirteen years he resided at his
farm, called Slighhouses, and visited Edinburgh only occasionally. Initially unsure of his decision to become a farmer, he soon
devoted all his energies to farming and related scientific experiments. This was unquestionably the most creative period of Hutton's life, comparable to Charles Darwin's five years aboard the
Beagle. When he packed up for Edinburgh in 1767, James Hutton was recognized as one of the leading mineralogists in Scotland, and he had quietly begun forming his theory of the earth.

Ironically, erosion, evident in so many parts of Scotland and the essential starting point for Hutton's theory, is not very obvious in the region around Slighhouses. It is a testament to Hutton's skills of observation that he properly assessed its power not by watching storm waves decimate the North Sea coast but by watching his soil wash away.

Slighhouses was located in the part of Scotland known as the Borders. The name reflected the history; the countryside bordered England and had changed hands many times in the past. Even before there was an England or a Scotland, the area was battled over by the Romans and the Picts. Hutton's land was not even 10 miles north of the border with England. The farm was situated in a section of the Borders known as the Merse, an ancient term that some believe means marsh, as the low-lying area may well once have been. The Merse consists of essentially flat land with gently rolling hills, which, like the hills of Edinburgh, were underlaid by the products of volcanic activity hundreds of millions of years ago. The numerous streams and creeks, called *burns* and *waters* respectively, flowing eastward to the North Sea helped to make the Merse one of the best farming regions in Scotland.

The area reminds an observer of southwest England, the region that Thomas Hardy wrote about, and the patterns of agricultural life in this part of Scotland in the eighteenth century were similar to those of Hardy's Dorset and Devon. The spring planting season and the fall harvest were intensely busy times, but the summer growing season and the dreary winters, during which there was little to do besides feed the animals, dominated. The farms were widely separated, and it was common to see no one but your own farmhands for days on end. Markets were held but once a week, and even then one saw only the same faces week after week in this sparsely populated area. Only the major holidays would have drawn out the boisterous crowds. The overall rhythm of life required a major adjustment for the urban born and bred Hutton.

THE TOPOGRAPHY OF THE BORDERS was nothing like that of the spectacularly rugged Highlands of northern Scotland, or even of the bumpy land surrounding Edinburgh and the Firth of Forth. But what the region lacked in dramatic features it made up for in vistas. Slighhouses itself was on gently sloping land, the fields rising toward the north. The elevation gave Hutton an uninterrupted view of vivid green fields for miles to the south, literally right from the stoop of his front door.

On a clear day, Hutton could stand at that door and look to the southeast across 12 miles of farmland and almost see the largest city in his region, Berwick (pronounced "Berick"), where the Tweed River runs into the North Sea. Berwick was fought over many times by the English and Scots, and in Hutton's day it was the northernmost city in England, as it remains today. It was a walled town during the medieval period, when it was Scotland's chief port. Because it was such an inviting target whenever the English and Scots were at war with one another, Berwick eventually lost its status as a key commercial hub. By the mid-1700s, though, having been unthreatened for generations, it was once again thriving and served as the chief port for northern England and southern Scotland. The streets of the town flowed downhill from the protective walls to the north bank of the wide Tweed. There were no village greens or parks to distract its 8,000 inhabitants from business. Yet, right before Hutton took up residence at his farm, the city had finished constructing a new town hall; this building immediately served as the main meeting place during market days. It seems likely that James Hutton came to Berwick whenever he needed to buy or ship anything substantial, such as new farm equipment.

The only other town of consequence was the southernmost Scottish port, Eyemouth, which lay about 9 miles east of Slighhouses and 8 miles north of Berwick. It got its name from the stream called the Eye Water, which flowed into the North Sea right in town. Eyemouth was much smaller than Berwick, and not nearly as active a port, but being Scottish, it was perhaps a more hospitable place for Hutton.

The rest of the Borders consisted almost wholly of various farms. The few towns, connected to one another by narrow dirt roads, were tiny, merely collections of a dozen or so sturdy homes. Two deserve specific mention. Just 5 miles from Slighhouses, and clearly visible from Hutton's front door, was Chirnside. This quaint village was built atop a 400-foot-high ridge, one street running east to west along the ridge, and the other one dropping straight downhill and south to the parish church, or kirk, below. The church was the largest in the district, parts of it dating from the twelfth century, and it was the closest to Slighhouses. The large public house, in the middle of town at the top of the ridge, would have been the center of activity on market days.

The other town that was part of Hutton's world was Duns, a market town only about 4 miles as the crow flies, and 6 miles by road, to the southwest of Slighhouses. Hutton later told his friends that his farm was near Duns, so this was most likely the town he viewed as home. Duns was very old, dating from the thirteenth century, and was built up around Duns Castle. The noble who founded the town maintained a large woods for hunting, and the forest around Duns to this day remains one of the few in southern Scotland, so intensively farmed is the land.

THOUGH HUTTON TRAVELED TO each of these villages and towns and did business in them, he spent the vast majority of his time on his farm. Slighhouses became Hutton property in 1713, when John Hutton, James's uncle, bought the 140-acre farm from a John Renton. William Hutton acquired it in 1718.

The grounds sloped up gently from south to north, the lowest point being about 300 feet above sea level, the highest about 450 feet. The house was situated in the middle of the tract. The northern border of the farm was on the edge of an upland; the southern border descended in the direction of the White Adder Water, a creek that formed the southern boundary of the neighboring farm. The White Adder is one of the main tributaries of the Tweed River, which it joined 3 miles from Berwick. Hutton's land was blessed with plenty of flowing water. There was Lintlaw Burn, Fosterland Burn, and at least five other streams that did not rate names. Evidence of Scotland's violent past was found nearby. About a mile from the house, in the upland, were the ruins of a small castle, called Bunkle Castle. Beyond the ruins were the remains of an ancient earthworks and castle, probably built by the Romans during their brief occupation of this part of Scotland in the first century A.D.

Hutton's house (which is still standing) was a standard two-story building with a slate roof—what today we call a Colonial. The main structure was built in the early 1700s, either by Renton or one of the Hutton brothers, and they had probably added on to an even older dwelling. The house had a very sturdy exterior, with what today looks like a stucco coating made of sand and pebbles. Hutton's home consisted of a bottom floor with two rooms, a parlor and the kitchen/eating area, and a second floor with two rooms, one being the bedroom. There was one simple entrance, a recessed, unadorned narrow door, and there were windows in the front and rear of each of the four rooms. As far as we know, Hutton lived all thirteen years at Slighhouses by himself, though he may have had a servant or two.

A second piece of property was also part of the Hutton holdings. In 1710, eight years before he bought Slighhouses, William Hutton purchased what is now called a hill farm. Named Nether Monynut, it consisted of 590 acres, a huge tract. Located in the Lammermuir Hills and reaching heights of 1,000 feet above sea level, this rocky, hilly land was never meant for cultivation; it was intended for grazing cattle and sheep. The hill farm was located about 8 miles northwest of Slighhouses, following a path alongside a creek called the Monynut Water. Hutton probably kept many of his cattle and sheep there most of the time, and then once a year herded some to Slighhouses to be fattened up and then sold at market.

WE KNOW THAT JAMES TOOK HIS time before finally settling at Slighhouses in 1754, a full five years after he finished his medical degree at the University of Leyden. Two letters written by Hutton to two friends in 1755, only a year after his arrival, suggest that he was very unhappy, perhaps living there under some form of coercion. In one he states: "This squeamish homebred stomach of mine an't truly reconciled to the bitter pill o' disappointment." There is a hint in these letters that he was heartbroken over the end of a relationship, which may have affected the mood and tone of the letters. To George Clerk-Maxwell (1715–1784; whose great-great grandson would be James Clerk-Maxwell, the discoverer of electromagnetism) he wrote, "I don't let any of the fair kind of creatures know of my distress; it would kittle the malicious corner of their hearts to hear the afflictions of a hardened wretch whom they could never make to groan." In the same letter he went on to say, "O if the ladies were but capable of loving us men with half the affection that I have toward the cows and calfies that happen to be under my nurture and admonition, what a happy world we should have!" In the second extant letter, this one to his lawyer in Edinburgh, John Bell, he alludes to no longer being socially active: "They had me at a feast of Baal in Eyemouth where was an honest sow roasted i' the gut so we had a dish of surprised pig and I did eat thereof; they led me up into the dance, but I will enter no more into their high places."

These are the only letters that remain from Hutton's years at Slighhouses, but it appears that shortly after they were penned, Hutton shook off his ill temper and became quite focused on the work at hand. He became an extraordinarily industrious farmer with a penchant for conducting experiments.

This was a period of rapid agricultural innovation in many parts of Europe, especially in England, and Hutton was determined to bring the lessons he had learned there to his native Scotland. He was remembered in the Borders region for introducing several new methods, which were then widely copied. First, he stopped using the traditional—and backbreaking—form of tillage, which was called *run-rig*. Run-rigs were high, straight mounds that were dug with the heavy and cumbersome Scots plough. They could be from 10 to 20 feet wide and several hundred yards long. The practice had begun hundreds of years earlier to provide drainage and to protect the crops from heavy rains. But the method was counterproductive; by loosening and then exposing so much soil to the elements, the losses to erosion were extraordinary. Instead of creating run-rigs, Hutton first enclosed his fields with low stone walls, and then added drainage ditches.

Hutton's second innovation was his use of a radically different plow, the Suffolk plough, which he had seen used during his stay in Norfolk. A Suffolk plough, light and well designed, was equipped with steel blades and required just two horses harnessed abreast to pull, and but one man to control. The Scots plough, used throughout Scotland at the time, was large, heavy, and made completely of wood. It required a team of at least six horses or oxen, and three men to handle. There is evidence that Hutton invited his neighbors to his fields to watch the new plough in action in the hope that they would be as impressed as he had been. Indeed, they were, and within a few years, Suffolk ploughs were in wide use in the Borders.

Hutton's method of crop rotation was also original and highly refined, and it, too, was copied by his neighbors. He rotated wheat, turnips, and barley, in that order. The turnips were planted late, in May, allowing that field to lie fallow for six months after harvest. When a field was lying fallow, he would graze cattle on it, for both the manure (a fertilizer) and the cleaning of crop waste that the animals ensured by their constant eating. The grasses with which he covered his turnip and barley fields kept a devastating turnip parasite at bay and made the barley field ready for grazing cattle immediately after harvest. Finally, Hutton was reputed to have kept his fields remarkably well weeded and clear, not because he was unusually neat, but because he realized that a clean field was best for the crops.

In addition to farming, the other main activity at Slighhouses was the raising of cattle and sheep. Though he slaughtered some cattle for beef, Hutton was more interested in the dairy goods they provided. In fact, one of the first things he did after arriving at Slighhouses was to build a dairy shed for the animals. The sheep, of course, provided meat and wool.

The wheat, turnips, barley, cattle, and sheep ensured one more thing at the farm—the presence of other people. Shortly after settling in the Borders, Hutton went to Suffolk to recruit a plowman, who resided with him for several years. The plowman supervised several farmworkers, most of whom probably lived in cottages on the farm. These were the farmhands who handled the daily chores of taking care of the animals and maintaining the equipment. During the spring planting season and fall harvest, dozens of extra workers from the surrounding towns would have been hired. In addition, the sheep would have required as many as twenty shearers during shearing season, which was in late spring. Most of the shearers probably lived in Chirnside and Duns, and they went from farm to farm performing their services.

James Hutton succeeded in turning his farm into a modern one, a model for the region. By the late 1750s, the hard work of restructuring the farm was finished. Hutton might have slacked off, but instead went to work on experiments that he had started soon after settling at Slighhouses. Using some of his sheds as labs, he investigated all kinds of problems that afflicted Scottish farmers. To assess the role of light and darkness on crops, he grew plants with calibrated amounts of sunshine. Organic and inorganic fertilizers were also the focus of his attention. One of his experiments dealt with isolating calcium carbonate in marl, an impure form of limestone, in an effort to help farmers put alkali into their fields and so improve the fertility of the soil. He also developed methods to eradicate smut, a devastating crop disease common in wet climates.

Though Hutton was immersed in the business of agriculture after his arrival in the Borders region, he clearly found time to pursue his geological and mineralogical interests, too. In 1764, he accompanied his friend George Clerk-Maxwell on a tour of the Highlands. Clerk-Maxwell was a member of the Commission for the Forfeited Annexed Estates. The government had formed this commission to assess and then sell the land confiscated from the clan chiefs following the rebellion of Bonnie Prince Charlie. Now known as a skilled mineralogist, Hutton's role was to help Clerk-Maxwell calculate the value of the land, both for farming potential and for its mineral holdings. The twosome made an arch across

the Highlands, moving from west to east (visiting Creiff, Dalwhin-
nie, Fort Augustus, Inverness, East-Ross, Caithness, Aberdeen,
then back to Edinburgh).

This trip marked the first time that Hutton had seen the
Highlands. An area of spectacular beauty, its high rugged mountains formed backdrops for lochs and waterways of all types.
Hutton would have viewed what is now called the Great Glen
Fault, the unique, arrow-straight line of thin lakes that marks the
boundary between the upper Highlands and the central Highlands. For someone recently interested in the study of the earth,
this trip to the Highlands would have been inspiring indeed.

JOHN PLAYFAIR, IN HIS *Life of Dr. Hutton,* muses that "it would
be desirable to trace the progress on an author's mind in the formation of a system where so many new and enlarged views of
nature occur, and where so much originality is displayed. On
this subject, however, Dr. Hutton's papers do not afford so
much information as might be wished for, though something
may be learnt from a few sketches of an Essay on the Natural
History of the Earth, evidently written at a very early period, and
intended, it would seem, for parts of an extensive work." The
essay that Playfair refers to represents Hutton's first musings on
his theory. It was probably written immediately after the 1764
excursion, although it has unfortunately been lost.

The two key tenets found in Hutton's essay were, one, that
most rocks are made up of eroded material (that is, sedimentary
rocks) and, two, that all surfaces on the earth are subject to constant erosion. As Playfair stated, "They were neither of them,

even at that time, entirely new propositions, though in the conduct of the investigation, and in the use made of them, a great deal of originality was displayed." It was seeing the two sides of erosion, and perceiving that they represented a cycle, that made Hutton's realization important and seminal. According to Playfair, who used expressions reflecting his and Hutton's belief in a supreme deity, Hutton understood "that, as the present continents are composed from the waste of more ancient land, so, from the destruction of them, future continents may be destined to arise. . . . Thus he arrived at the new and sublime conclusion, which represents nature as having provided for a constant succession of land on the surface of the earth, according to a plan having no natural termination, but calculated to endure as long as those beneficent purposes, for which the whole is destined, shall continue to exist." As Playfair makes clear, in this early essay, dating from the mid-1760s, Hutton was already arguing that the earth was ancient, and that it would continue with its cycle of destruction and rejuvenation until the Creator Himself brought the mechanism to an end.

Stephen Jay Gould has referred to Hutton's realization as "the paradox of the soil." Erosion is necessary to form soil, a key to the survival of all humans, but it also destroys the very soil that it has formed. Without a mechanism for its restoration, the land would quickly become uninhabitable, which the deist Hutton believed a benevolent Creator-God would not allow.

THE IRRESISTIBLE QUESTION IS THIS: When did James Hutton, this clever farmer, start developing his theory of the earth, however fitfully? The answer is that we do not know; Hutton kept no diary, and the sparse correspondence that survives makes no mention of a starting point. We do know that Hutton was paying attention to minerals and rock formations during his stay in Norfolk.

One source of insight into erosion, in addition to his disappearing soil, may have been the ruins of Bunkle Castle and the old Roman earthworks near his house. The 500-year-old foundation stones of the castle would have been transformed into smooth, rounded protrusions by the Scottish rains and wind. The earthen walls of the Roman structure would have been even more severely damaged. Hutton recognized that even though erosion was constantly occurring, it nonetheless operated quite slowly.

As for sedimentary rocks, they were all around him. We know that one of the first things that Hutton did at Slighhouses was to enclose his fields with low stone walls. "A cursed country where one has to shape everything out of a block & to block everything out of a rock. . . . I find myself already more than half transformed in to a brute," he wrote in 1755. Because he worked on the land day after day, there would have been no other time during which he handled rocks so intensively. At the same time he was gathering and stacking sandstone blocks, he would have been noticing his soil leeching away and ending up in the numerous streams on his property. Watching the fast-flowing water wash his soil downstream, Hutton would have imagined the journey from Lintlaw Creek to the White Adder, which flowed into the Tweed, which then emptied into the North Sea. Perhaps this freethinker took note. Perhaps he realized that the erosion he saw on his land every day put grains of dirt and sand into streams, which flowed into rivers, which then flowed into seas, where the grains settled to become sediments, which would eventually become sedimentary rocks. And he was the first to realize that these future sedimentary rocks would one day become new dry land and replace the fields he was currently farming.

In 1767, his farm in good order and his theory of the earth beginning to coalesce, James Hutton packed his bags once again, and finally headed back to Edinburgh. Unbeknownst to him, he was going back to participate in one of the most remarkable periods of intellectual discovery to have occured in one place in the Western world.

The Athens of the North

No place in the world can pretend to competition with Edinburgh.

Thomas Jefferson, 1789

T HE EDINBURGH TO WHICH James Hutton returned in late 1767 was quite different from the one he had abandoned twenty years earlier to continue his studies in Paris. Most of the changes could be traced back to the rebellion of Bonnie Prince Charlie in 1745–1746 and its aftermath. Also, they were chiefly the direct or indirect work of one man, Lord Provost George Drummond, who died at the age of seventy-nine the year before Hutton's return, after having led the city for twenty years. Drummond had co-organized the defense of Edinburgh, along with Colin Maclaurin, in September 1745, when it became clear that the government's soldiers would not return in time to protect the city. He was appalled by the cowardice shown by his fellow citizens and incensed by the manifestation of rampant Jacobism.

When he was made Lord Provost in early 1746, he did not rest until all Jacobite supporters were essentially run out of town, either through moral suasion or imprisonment (for aiding and abetting an enemy). While the Duke of Cumberland was "harrying the glens" and eliminating Jacobism in the Highlands, Drummond was harrying Edinburgh with essentially the same goal, though the rule of law prevailed. The purging of the Jacobites largely drove the old aristocracy from the city so that whatever class distinctions had once existed were now gone. Edinburgh became dominated by pro-business, pro-union (with England) Whigs—progressive merchants, solicitors, local government officials, and scholars.

A similar process was taking place in the Presbyterian Church, spurred by a faction known as the Moderate Party. Led by William Robertson, historian and soon-to-be principal of the university, the Moderates represented a wing of the church that saw "industry, knowledge, and humanity linked together by an indissoluble chain." They believed that it was possible to harmonize the goals of the Whigs and a modern commercial society with those of the church. The leaders were young ministers, many of whom had been part of Drummond and Maclaurin's student brigade.

Thus, in the years after the 45, the city saw many tensions evaporate. The Jacobite threat was over forever, which meant that Scotland would never again have to worry about armed aggression; future wars would be England's concern. A large portion of the city's population, now about 50,000, shared the same basic attitudes toward commerce and progress. And the previously strict and conservative Presbyterian Church became more tolerant, especially toward the goals of the Whigs.

George Drummond also strove to further improve the university and medical school, two institutions that he doted on like a parent. Thanks to his care, by 1767 the most able professors from Scotland's other universities had been lured away from their home schools and hired by one of Edinburgh's. As a result, both were now superior institutions, attracting students from all over Europe and making Edinburgh a Mecca for scholars. This was particularly true for the sciences. The spirit of Newton and Maclaurin was maintained by Matthew Stewart, one of the brightest mathematicians in Britain. The two leading chemistry professors in Scotland, William Cullen and Joseph Black, had been brought from the University of Glasgow. Learned social clubs, of which only a few had existed before 1745, were now ubiquitous, giving scholars a place to congregate and discuss ideas nearly every day.

All these positive developments led to continued population growth. Realizing that the physical constraints of the 600-year-old city hampered its ability to absorb more people, Drummond had finally convinced Edinburgh's leaders to break out of the ancient confines and construct a completely new section. Called, originally enough, the New Town, this jewel of urban planning would be built to the north, between the old town and the port of Leith. When finished, it would provide the residents with all the space they would ever need, and would allow Edinburgh to grow to its natural size, no longer trapped behind the Flodden Wall. The final design was officially accepted in the spring of

1767, and construction had begun just as Hutton moved back that fall. The plan called for the filling in of Nor' Loch, the man-made lake at the north base of Castle Rock, which served as a moat in the earliest days, and was now an open cesspool. Drummond's singular vision would be a complete departure from the old town—laid out as a grid, it would feature dozens of wide streets and two large parks.

There was one other notable development, which was less tangible than the stones and mortar of New Town, but significant nonetheless. Edinburgh had just produced its first genuinely world-famous personality, the philosopher and historian David Hume (1711–1776). Born and educated in Edinburgh, Hume wrote his most important book of philosophy before this period; the massive *A Treatise of Human Nature* was published in 1739–1740. The main theme of Hume's philosophy—an idea that would eventually embed itself into the culture of late eighteenth-century Europe and North America—was that humans were largely controlled by their passions, not their rationally trained minds, and that this was natural and therefore good. "Reason is, and ought to be, the slave of the passions," he wrote. Thousands of years of religious and philosophical thought, from the Greeks to Rousseau, had sought to elevate human beings above the beasts of the jungle because of their reason and their ability to overcome base passions. Hume essentially said that humans were still beasts. Moreover, he argued, only by accepting the reality that humans are motivated by self-interest and passions would governments be able to create proper and effective civil institutions. Hume was also an avowed atheist and critical of all religions, mainly because they sought to curb human passions. To Hume's chagrin, his *Treatise*, written over ten years, went largely unread, at least initially.

Hume was not present in Edinburgh during the 45, but he became galvanized by it all the same, and soon after decided to write a history of the Stuarts. Taking advantage of his position as Keeper of the Advocates Library, the finest collection of books and documents in Edinburgh, Hume ultimately wrote a six-volume history of England and Scotland, published from 1754 to 1762. He used it to demonstrate how the human race, motivated by passions and self-interests, could still experience progress. *The History of England* became a best-seller—it would go through multiple editions over the next fifteen years—and made him a wealthy man. Hume went from living modestly to spending lavishly. He took advantage of his success and reissued many of his philosophy books, which now became widely recognized as profoundly important works. They, too, sold well. Collectively, these publications sought to create a "science of man" derived from Newtonian principles and applied to human behavior. Hume's ideas would exert a great deal of influence on Thomas Jefferson, Benjamin Franklin, and the rest of the founding fathers of the United States.

A naturally social man, Hume had openly sought fame and celebrity all his life, and now he finally had it. From 1763 to 1766, Hume was the secretary and then chargé d'affaires at the British embassy in Paris. He became the toast of the town, befriending Denis Diderot, Jean d'Alembert, and Jean-Jacques Rousseau. The British government transferred him to London in 1767, where he served as the undersecretary of state for the next two years. Then, in 1769 and at the age of fifty-eight, David Hume retired to Edinburgh, where he would be the de facto leader of the scholarly community until his death in 1776.

In the two decades of James Hutton's absence, Edinburgh had metamorphosed from a medieval city to one of the most modern in Europe. In 1767, it was squarely on the map of progressive Europe and no longer a backwater town in a lawless country known only for its inhospitable climate.

ANOTHER FACTOR THAT CONTRIBUTED to the energy in Edinburgh was the thriving economy of the city and the country as a whole. The textile, fishing, and banking industries had each experienced recent productive innovations. And there were expectations for even greater prosperity now that construction was about to begin on the Forth and Clyde Canal, the water passage that would connect the Firth of Forth with the Firth of Clyde, and thus the twin cities of Scotland: Edinburgh and Glasgow.

The canal was among the reasons why James Hutton returned to Edinburgh when he did. Canal building was the rage in mid-eighteenth-century Europe, as new tools, techniques, and materials finally allowed for the long-desired connections between large commercial centers. Though the Forth and Clyde was not the first canal in the British Isles, at a planned 38 miles long it was among the most ambitious. Surveying had begun in 1762, and the actual digging would begin in June 1768 (the canal would be completed in 1790). An organization called the Forth and Clyde Navigation Company had parliamentary consent to supervise the enterprise, and it appointed a Scottish committee of management to oversee the construction. The committee consisted of nine men, one being George Clerk-Maxwell, with whom Hutton had toured the Highlands in 1764. Maxwell wanted Hutton on the canal committee for the same reason he had wanted him to examine the forfeited Highland estates—he was a talented mineralogist and the canal committee would constantly be facing such geological issues as determining the ideal path along which to cut the canal, deciding on the proper prices to pay for land, and choosing the quarries that would supply the needed materials.

Hutton became a member of the canal's committee of nine in the last part of 1767, and he stayed on until 1775. By then the committee, which was originally based in Edinburgh, had moved to Glasgow, since the main construction of the canal was now nearing the Clyde. The records of the canal company indicate that Hutton attended over eighty often-lengthy meetings in seven years, and that he was heavily engaged in day-to-day decisions in the early years of his participation. Being able to observe much of the actual digging added to Hutton's already extraordinary knowledge of the geology of Scotland.

JAMES HUTTON'S INVOLVEMENT IN the construction of the Forth and Clyde Canal was the only formal obligation he would have for the rest of his life. He likely remained involved with the Davie/Hutton Sal Ammoniac chemical works because in 1765 the partnership was formalized, but there is no indication that he had any essential responsibilities or that he spent much time there (although he did have materials sent to him at the chemical works address). Indeed, Hutton had no need to work. He returned to Edinburgh a wealthy man (he was among the original investors in the canal, having paid 500 pounds to acquire five shares). Moreover, he would continue to collect a steady stream of income from the Davie/Hutton works, from Slighhouses, which he rented out, and from various properties in Edinburgh: "The rents, mails, and duties . . . of the houses, shops and others in the Town of Edinburgh pertaining and belonging to me."

Hutton remained single, so he had no family obligations (it appears that he sent money to his son in London, but there is no evidence that he ever saw him). He did, however, live with his three sisters, who never married, in the house that he built in 1770. It is probable that his mother was now deceased, otherwise she would have moved in with him, too. A detailed description of the house was left by a relative of Hutton's, Sir James Crichton-Browne (1840–1938), who was born in the house forty-three years after Hutton's death:

(The house) was (in) a picturesque corner of Old Edinburgh. It stood on a ridge about 120 feet above the south back of the Canongate, and was a cul de sac, approached from St. Leonards by a narrow pathway and a curious old arched passage, and from the south back of the Canongate by a long flight of broad but much dilapidated steps. There were only seven or eight houses in St. John's Hill, each detached and standing in its own bit of ground, and shadowed by its own trees. The house at St. John's Hill, standing back from the road, and overshadowed by trees, was approached by a gateway and a short walk, and was very much like the houses of the well-to-do in Edinburgh in those days. On the right, on entering, was a long dining room, rarely used, to the left, a small parlour that was the family rendezvous, and upstairs there was a long fusty drawing-room, only opened on state occasions, and a number of bedrooms, all stiffly furnished, and with four-poster beds. At the back of the house was a green on which we putted.

When Hutton lived there, at least one of the rooms served as his laboratory. Soon after the house was built, a visitor wrote: "His study is so full of fossils and chemical apparatus of various kinds that there is barely room to sit down."

St. John's Hill was a curious location for the house. At a time when New Town was under way, and when David Hume and others would soon be building new and spacious houses there, Hutton chose to build in the old town. The plot was just outside the Flodden Wall, toward the end of the Royal Mile near the Palace of Holyroodhouse. It was conveniently close to the University of Edinburgh and the infirmary. The house was also within a couple of blocks of the Davie/Hutton chemical works. But no doubt the main reason he built there was that the spot gave him an unblocked view of Arthur's Seat, the massive mound at the eastern end of the city.

Arthur's Seat is a highly unusual geologic formation, and it was particularly inspirational for Hutton. There was no way for the doctor to grasp all the geologic phenomena revealed there, but he must have sensed a fascinating history. Indeed, this one spot embodied the deep past of the Edinburgh region: 350 million years ago, a volcano erupted, creating a volcanic cone; then the area was submerged under an ancient sea, the waters eventually receding (or the land rising), leaving sedimentary rocks around the volcanic cone; then followed earthquakes and mountain-forming pressure that further raised and distorted the mound; and finally an ice-age glacier overran it all. However, eventually Hutton did recognize that the oddest feature, the wall of exposed dark rocks that traverses the side of the hill, called Salisbury Crags, was different from the other rocks on Arthur's Seat. He alone realized that the Crags were younger rocks than the strata around them. This realization became important later, as he pondered the significance of subterranean heat and igneous rocks. Arthur's Seat offered daily lessons for the philosopher.

Because Hutton had no worries or obligations, each day was his own to do with as he pleased. According to Playfair, James Hutton's consistent daily habits allowed him to remain remarkably focused on his geological inquiries, which now completely dominated his thoughts. He was a late sleeper, but once out of bed he went directly to his study and began working. He ate his small midday meal quickly and alone. After eating, he would return to his studies for a few more hours, and then go for a long walk, often along the paths up and down Arthur's Seat, weather permitting. He read a great deal, primarily natural histories and travelogues. These were used to augment his own observations and verify his theories. He spent every evening with friends, either at home or at a tavern: "No professional, and rarely any domestic arrangement, interrupted this uniform course of life, so that his time was wholly divided between the pursuits of science and the conversation of his friends."

Most of Hutton's companions were other natural philosophers. But this was not just any collection of academics. Rather, the scholars who were part of James Hutton's circle in Edinburgh were so freethinking, so forward looking, so productive, and so prolific that collectively history remembers them as members of the Scottish Enlightenment. It was the environment these thinkers created—noted for constant personal interaction and the debate of new ideas—and the specific teachings that they shared that helped Hutton take what was a flash of insight formed at Slighhouses and turn it into something far more complete. The Scottish Enlightenment essentially served as the incubator for Hutton's nascent idea and gave it the support and protection it needed to mature as a fully realized, rigorous, and robust theory. The scholars of the Enlightenment would also create the institution through which Hutton would announce his theory to the world in 1785.

THE SCOTTISH ENLIGHTENMENT was an intellectual movement that complemented the Whig regime in the city. It celebrated progressive ideas and witnessed significant contributions in fields as diverse as geology, mineralogy, chemistry, medicine, political economy, history, philosophy, architecture, poetry, and portraiture. If there was a unifying theme or philosophy, it was that the "improvement" of the natural world—by means of understanding and controlling it—was fundamentally good and proper. Related to this was the idea that Newton-inspired natural laws could and should be applied to many phenomena, such as human nature and human history. Immanuel Kant's characterization for the Enlightenment on the Continent also described the Scottish version: "Dare to know."

A group of native Scots, nearly all of them educated at Scottish universities and most living within blocks of one another in Edinburgh, along with regular visitors from Glasgow and other nearby towns, made up the cast of enlightened scholars. They were primarily university professors, ministers, and lawyers/solicitors; as one historian calls them, the "teachers, preachers, and pleaders" of the city. The Scottish Enlightenment spanned two generations: the more influential group was born before 1740, and the second wave was the following generation.

A short list of the most active participants includes Adam Ferguson (1723–1816), considered the founder of sociology because of his book *Essay on the History of Civil Society* (1768); William Robertson (1721–1793), one of the founders of modern historical research and noted for his *History of Scotland* (1759); William Smellie (1740–1795), the printer and publisher who compiled and edited the first edition of *The Encyclopedia Britannica* (published in installments from 1768 to 1771); William Cullen (1710–1790), one of the leading early medical researchers and chemists of the time; Sir John Clerk of Eldin (1728–1812), who became the Clausewitz of naval warfare because of his book *An Essay on Naval Tactics* (1790–1797); Robert Adam (1728–1792), the influential architect; Robert Burns (1759–1796), the great poet; and Sir Walter Scott

(1771–1832), who "invented" the historical novel and who came of age during this vibrant period. Onto the list should also be added two men who never lived in Edinburgh but who visited and maintained an active correspondence with the scholars there: Ben Franklin (1706–1790), the statesman and talented polymath who discovered electricity; and Erasmus Darwin (1731–1802), Charles Darwin's grandfather and the author of a precursor theory of evolution. John Playfair (1748–1819) and James Hall (1761–1832) were key figures as well.

But beyond this impressive group, five individuals made contributions so monumental that they still reverberate to this day, over 200 years after their deaths. David Hume developed and published his influential philosophy of human nature, which argued for the primacy of man's passions. Adam Smith (1723–1790), one of Hume's best friends, used parts of Hume's philosophy while writing *The Wealth of Nations,* the book that started the field of economics and allowed governments to finally understand the effects of laws on their nation's economy. Joseph Black (1728–1799) isolated carbon dioxide, thus discovering that the atmosphere was made up of a mix of gases, and inspired Antoine Lavoisier, the founder of modern chemistry. James Watt (1736–1819), who worked in Black's lab, went on to invent the practical steam engine. Finally, there was James Hutton, the father of geology and the discoverer of the antiquity of the earth.

Hutton's relationship with Hume is unknown, for there is no record to show that the two ever met (it is difficult to imagine that they did not know each other, however, given that they shared a mutual friend in Black). The others were all Hutton's friends,

Joseph Black being the most important. It was Black who shared his love of chemistry and helped Hutton navigate the social and intellectual network that was the Scottish Enlightenment.

Based on the observations left by those who interacted with Hutton after he moved back to Edinburgh, the doctor-turned farmer-turned natural philosopher must have been regarded as a curiosity by those who encountered him. As one mineralogist wrote to another after meeting Hutton during a visit to Edinburgh, "Dr. Hutton is the oddity you described, but a mighty good sort of man." Just returned from his farm in the Borders, surrounded by rocks and chemicals in his overcrowded flat, he would have been hard to take seriously. Even his attire was off; he was described as being careless in what he wore, and "often found in direct collision" with the accepted fashions. But not Joseph Black. An internationally known scientist before the end of the 1750s, probably the brightest of the many stars at the University of Edinburgh, and connected to every prominent citizen in the city, Black was the very essence of an insider. In fact, one historian of the period calls Black the éminence grise of the era. Black quickly realized the unique talents possessed by Hutton, and if Black said he was good company, then just about everyone else in Edinburgh soon felt the same way.

ALTHOUGH OBSERVERS OF THE Scottish Enlightenment often treat David Hume as the preeminent thinker of that period, Joseph Black (he later became David Hume's doctor) was the first to make a profound discovery. In the 1750s, while still a medical student at the University of Edinburgh, Black burst onto the scientific scene when he discovered carbon dioxide, the first-ever instance of isolating one of the gases in the atmosphere; indeed, before Black's discovery, no one had imagined that the atmosphere was a mixture of individual gases.

The path to this discovery was particularly significant for Hutton because it involved a common mineral that would later serve as proof of his theory of the earth. The experiment started out modestly enough. Attempting to help resolve a conflict between two of his professors, Black sought to discern the most efficient way to dissolve urine stones, a common affliction in the eighteenth century. He chose to experiment with magnesia alba, a type of limestone (limestone was already recognized for having certain medicinal characteristics). While making some simple measurements, he stumbled across an amazing thing. When the magnesia alba was heated, it lost 40 percent of its weight (it is hard to imagine today, but Black was among the first chemists to use a scale, a tool almost as important as the microscope). Black applied the term *fixed air* to the weight that was lost. Black's next step was to investigate the properties of the fixed air that had been released. If trapped and unmixed with the surrounding air, this fixed air killed living things, such as mice. Since this toxic fixed air was clearly in the atmosphere (the process of burning limestone is common enough), it followed that there must be other components of the atmosphere that acted to dilute the toxic properties. The realization that the atmosphere was made up of a mix of discrete gases was a revelation, setting the stage for Lavoisier to propose the oxygen theory in 1774 and found modern chemistry.

The paper that formally described Black's discovery, "Experiments on Magnesia Alba, Quicklime, and other Alcaline Substances," published in 1756, was immediately recognized as seminal. Black, only twenty-eight years old and just out of medical school, was quickly deemed the leading chemist of his generation. Over the next two years, Black made another important discovery: He deduced the existence of latent heat. When a compound is in the process of changing states (for example, liquid water converting to steam), the compound continues to absorb heat, even though its temperature remains unchanged. While exploring the phenomenon of latent heat, Black came to understand the role that pressure played on heated substances. Water, for instance, maintains its liquid state under pressure when its absorbed heat would otherwise cause it to be converted to steam. Black's insights into heat and pressure would be vitally important for Hutton's work.

It is not known precisely when Black and Hutton first met, but shortly after Hutton moved to Edinburgh, the two scientists became fast friends. In 1771, Black wrote to James Watt, "I wish I could give you a dose now and then of my friend Hutton's company, it would do you a world of good." They would remain the closest of friends for the rest of their lives. Adam Ferguson, Black's contemporary biographer, wrote that Black's close friends were Cullen, Watt, Hume, Smith, Monro, and Clerk of Eldin, but that

at the head of either list, however, in respect to Black's habits of intimacy, ought, perhaps, to have been placed James Hut-

ton, who made up in physical speculation all that was wanting in any of the others. It may be difficult to say, whether the
characters of Black and Hutton, so often mentioned together,
were most to be remarked for resemblance or contrast. . . .
Black was serious, but not morose; Hutton playful, but not
petulant. The one never cracked a joke, the other never
uttered a sarcasm. Black was always on solid ground. . . .
Hutton, whether for pleasantry or serious reflection, could be
in the air.

It appears that they went to work on Hutton's mineralogical
research without delay, chemistry now being the key tool
through which Hutton pursued his inquiries. As Playfair said, it
was through "Chemistry . . . that he took his departure in the
circum-navigation both of the material and intellectual world."
What he and Black focused on was the largest conundrum that
Hutton unearthed at Slighhouses—the mystery of the mineralizing principle. Most mineralogists believed that all visible rocks—
granite, basalt, stratified rocks, marbles, and so forth—were
precipitates (mineral remnants) from the universal ocean. Hutton found this difficult to believe because he had observed
through his experiments that every possible substance appeared
in rocks, even substances that could not be dissolved by water. If
every rock on the surface of the earth had precipitated from
water, then water must be able to dissolve every substance found
in rocks. But this simply was not true. During the course of his
research with Black, Hutton resolved that heat from within the
earth was the only conceivable mineralizing principle. However,

it was crucial that heat be coupled with intense pressure, which always exists under the heavy sediments that rest on the ocean floor.[1] Over the next fifteen years, Hutton would take his early insights from Slighhouses (that erosion is acting on all materials at all times, and that the land we are living on is made of ancient eroded material), combine them with his realization about the power of heat and pressure, and develop them into a stunningly original theory.

AN EXTRAORDINARILY POPULAR TEACHER, Joseph Black had many outstanding students who went on to make significant intellectual contributions. His most famous young charge was easily James Watt.

Watt was hired as the "mathematical instrument maker to the university" of Glasgow in 1757, one year after Joseph Black started there as professor of chemistry (Black moved to the University of Edinburgh in 1766). He worked with both Black and John Robison, who was professor of physics at Glasgow and would also eventually move to the University of Edinburgh after Black. Watt was an astute young man who had a bent for both

---

[1] As mentioned earlier, geologists today understand that the mineralizing principle (the modern term is *lithification*) for sedimentary rocks is pressure coupled with the cementing that occurs when water and air are squeezed out of the material. Though Hutton was wrong about the need for heat for the formation of sedimentary rocks, he was the first to recognize the importance of pressure. Hutton, however, was right in believing that heat and pressure are the keys to the formation of metamorphic rocks, though in the late eighteenth century, metamorphic rocks were considered sedimentary.

science and engineering. In the winter of 1763–1764, he was asked to repair a working model of the Newcomen steam engine, which the university had purchased to further understand how it worked (the Newcomen engine was the first steam-powered engine; invented in 1712, it was very large and slow, and was mainly used in settings where its sluggish power made sense, such as driving a slow-acting pump in a mine, its first application). Watt was able to repair the model, but as he was fixing it he realized that it was a grossly inefficient contraption. The Newcomen engine had just one cylinder, to which the piston was attached; the heating and cooling of the single cylinder created the vacuum that caused the piston to rise and fall, and power whatever was attached to it. But the energy loss was tremendous, and the engine was plodding. Watt realized that if a second cylinder could be added that would quickly provide the cooling/heating action to the piston-cylinder, one could achieve significantly greater speed and productivity. This is precisely what he built: a two-cylinder, separate condenser, steam engine. He patented the idea in 1769.

Through Black, James Hutton and James Watt became close friends, too. Hutton accompanied Watt on his historic move from Glasgow to Birmingham to begin his partnership with Matthew Bolton, a pairing that would play a central role in the Industrial Revolution. Leaving Glasgow in the early summer of 1744, the two travelers reached Birmingham about two weeks later; Hutton then took his leave of the young inventor and spent the next several weeks exploring the geology of southwest England and Wales. Though this period is the only documented geologic excursion that Hutton embarked on from 1764 (when he visited the Highlands with Clerk-Maxwell) to 1785 (after his lectures), there were surely others. After the doctor returned to Edinburgh in the late summer of 1744, he claimed in a letter to Clerk-Maxwell that "I think I know pretty well now what England is made of except for Cornwall."

The other important introduction made by Black on Hutton's behalf was to Adam Smith. Smith was perhaps David Hume's best friend; they had known each other for twenty-five years when Hume died, and had conducted one of the most learned correspondences in intellectual history. David Hume fell ill in 1775 with what was either intestinal or stomach cancer. He died one year later, at his home in New Town. Joseph Black was Hume's doctor (Black was a medical doctor, as were many chemists at the time, but he had only a handful of patients), and he wrote to Smith many times during Hume's difficult last year to report on his health. As a result, the two developed a friendship.

Adam Smith had been a well-known philosopher and writer since the appearance of his first book, *The Theory of Moral Sentiments,* in 1759. He finished his second book, *An Inquiry into the Nature and Causes of the Wealth of Nations* (1776), several months before Hume's death. Smith had been working on it for years, so the book was widely anticipated. The response to its appearance was immediate and enthusiastic; the first printing of the book was sold out in six months in an era when books were usually printed just once. It was quickly reviewed in every major publication in England and Scotland, and overnight it achieved its status as one of the clearest, most useful, and most influential

books ever written, a reputation it still enjoys today. The goal of
the book was to analyze how national economies grow, and in so
doing it made at least five powerful pronouncements: that an
economy was a system (in the spirit of Newton), meaning an
action in one sector would have an impact in another sector; that
the growth of wealth meant the increased buying and selling of
goods, such as clothing, not the accumulation of money (gold or
silver); that the key to efficient buying and selling was free markets and an uncontrolled price system; that the key to growth was
innovation; and that the key to innovation was the division of
labor, which allowed for efficiency and the gaining of expertise.
All five of these precepts have stood the test of time. The book
had immediate practical influence, too. The British chancellor of
the exchequer proposed new tax initiatives for the national bud-
gets of 1777 and 1778 based on Smith's book. William Pitt the
Younger followed its teachings on free trade and unencumbered
markets when he was prime minister in the 1780s.

After the publication of *The Wealth of Nations,* Smith was
appointed one of the five commissioners on the Scotch Board
of Customs because the leaders of Edinburgh wanted to keep
their famous citizen in Scotland. Thus, Smith moved to Edinburgh for the first time in 1777 after having lived most of his
life in Kilkaldy (25 miles away) and Glasgow (45 miles away).
The work at the Board of Customs was easy and routine, but
Smith took the job seriously and rarely missed a day of work.
He told acquaintances that the job was very useful to his understanding of political economy, and future editions of *The
Wealth of Nations* (four more were published before he died)

were enriched with numerous examples from cases and disputes brought before the board.

He bought a spacious home soon after he arrived in the city, and brought his elderly mother, a cousin, and her nephew to live with him. The house was located just off High Street, near Holyroodhouse. It was only a couple of blocks from James Hutton's house, and Joseph Black lived nearby, too. The Custom House, where Smith worked, was also situated on High Street, about half a mile up the ridge, heading toward the castle. Smith's habits were rigid; he walked up High Street at the same time each morning on his way to work, and strolled back down in the late afternoon. He cut quite a figure. He always carried a cane, but never used it—rather he rested it on his shoulder "as a soldier carries his musket." He dressed well, but not extravagantly. Like Hutton, he struck observers with his eccentricities because he often talked to himself, and his head turned from side to side while he walked. One of his biographers commented, "Often, moreover, his lips would be moving all the while, and smiling in rapt conversation with invisible companions."

Like Hutton and Black, Adam Smith was unmarried. Though he was sociable and fond of company, accounts of the period indicate that he was shy and spoke up only if called upon. His biographer mentioned that "Smith's house was noted for its simple and unpretending hospitality. He liked to have his friends about him without the formality of an invitation, and few strangers of distinction visited Edinburgh without being entertained in Panmure (Smith's) House." Every Sunday night he hosted a dinner for friends and guests, and lively discussions were encouraged.

INFORMAL BUT REGULAR DINNER parties like Smith's were common in Edinburgh during the years of the Scottish Enlightenment, and they were one of the ways in which ideas were exchanged. A productive cross-fertilization often resulted. Allan Ramsay, a successful painter and the son of one of Scotland's most noted poets, once wrote in a letter: "By much drinking with David Hume and his associates, I have learnt to be very historical; and am nightly confirmed in the belief, that it is much easier to tell the How than the Why of any thing; and that it is moreover better suited to the state of man; who, we are all satisfied, from self-examination, is any thing rather than a rational animal."

Hutton and Black followed the practice as well. As one University of Edinburgh student wrote in the 1780s: "We are going to take a Christmas dinner with Dr. Black on Monday next where we meet a good number of students. The Dr. himself is so lazy he is obliged to get Dr. Hutton to be master of all the ceremonial part. The Dr. [Black] likes to sleep after dinner."

Social clubs were yet another means of intellectual exchange. Other cities had clubs, of course, but Edinburgh was known to have more of them and a greater diversity. One of the first ones founded was the Cape Club; a member summed up its raison d'être: "The purpose and intention of the Society was: after the business of the day was over to pass the evening socially with a set of select companions in an agreeable and at the same time a rational and frugal manner; for this purpose beer and porter were their liquors, from fourpence to sixpence each the extent of their usual expense, conversation and a song their amusement, gaming generally prohibited, and a freedom for each to come and to depart at their pleasure was always considered as essential to the constitution of the Society."

The clubs almost always met weekly in a public house or oyster cellar. The company often included both men and women, and distinguished out-of-towners were welcome. Many citizens of Glasgow regularly made the 45-mile, one-day journey. In addition to the Cape Club, there was the Boar Club (members were called "boars," the room they met in was called the "sty," etc.); the Mirror Club, which met to promote agricultural improvement; the Rankenian Club, which discussed philosophical issues; the Poker Club, which formed to lobby for a Scottish militia; and the Crochallan Fencilbles, which sought to encourage Scottish literature. The Select Society, started in 1754, became more ambitious by scheduling one of its members to present a formal speech, later to be discussed and critiqued by those assembled. As one member later recalled, "These convivial meetings frequently improved the member more by free conversation than the speeches in the Society. It was those meetings in particular that rubbed off all corners, as we call it, by collision, and made the literati of Edinburgh less captious and pedantic than they were elsewhere."

Though Black and Smith were members of several of these clubs, it does not appear that Hutton frequented them. But soon after Adam Smith settled in the city in 1777, he, Black, and Hutton founded their own club, known as the Oyster Club. Smith's biographer provided this description: "I have already mentioned his Sunday suppers, but beside these he founded, soon after settling in Edinburgh, in co-operation with the two friends who were his closest associates during the whole of this last period of his career—Black, the chemist, and Hutton, the geologist—a weekly dining club, which met every Friday at two o'clock in a tavern in the Grassmarket." Dr. Swediaur, a Paris physician who spent time in Edinburgh in 1784 conducting research with William Cullen, and was made a member of the Oyster Club during his stay, wrote to Jeremy Bentham: "We have a club here which consists of nothing but philosophers. Dr. Adam Smith, Cullen, Black, Mr. McGowan, etc., belong to it, and I am also a member of it. Thus I spend once a week in a most enlightened and agreeable, cheerful and social company." John Playfair left a similarly affectionate comment: "As all three possessed great talents, enlarged views, and extensive information, without any of the stateliness and formality which men of letters think it sometimes necessary to affect, as they were all three easily amused, and as the sincerity of their friendship had never been darkened by the least shade of envy, it would be hard to find an example where everything favorable to good society was more perfectly united and everything adverse more entirely excluded."

THE SPIRIT OF COLIN MACLAURIN had most tangibly lived on through the existence of Edinburgh's most formal learned society, the Philosophical Society, founded in 1737. Maclaurin felt that Edinburgh needed an answer to London's famous

Royal Society (to which he already belonged), so he convinced Alexander Monro, the first professor of the medical school, to broaden his Society for the Improvement of Medical Knowledge (started in 1731) into a general scientific organization. For many years after Maclaurin's death, the Philosophical Society remained active; Joseph Black's paper on magnesia alba and carbon dioxide was published in one of its three proceedings volumes. Hutton had joined the society soon after settling in Edinburgh, probably through Black's sponsorship. Though it never actually ceased to exist, by the late 1770s, it had become moribund.

In 1782, a group outside the university planned to petition the government in London to grant them a royal charter to found a natural history museum and sponsor lectures. John Walker (1731–1803), professor of natural history at the university and the curator of its natural history museum, objected vehemently to this proposal. He felt strongly that the new museum would dilute the university's holdings because they would be competing for the same specimens, archives, and artifacts. So Walker lobbied for the university to reach a compromise with the competing group.

The outcome of a convoluted negotiation was the founding of the Royal Society of Edinburgh, by royal charter, in May 1783. The new society was officially a merging of the Philosophical Society with parts of the university. Its purpose was to encourage and disseminate outstanding scholarship in the sciences, philosophy, history, and literature. Nearly all of the sixty fellows of the Philosophical Society became the founding fellows of the Royal Society. The roster was a "who's who" of the Enlightenment: Black, Smith, Cullen, Ferguson, Robertson, Robison, Colin Maclaurin's son John, and, of course, Hutton.

Sometime in 1784, an invitation was formally extended to Hutton to present to the society two lectures on his theory of the earth. This prestigious invitation would force Hutton to synthesize the work and reflection he had conducted over the past thirty years. And it would spark an intellectual revolution that would eventually lead to modern geology, evolutionary biology, and an understanding of the true age of the earth.

8

The Eureka Moments

Dr. Hutton made several excursions into different parts
of Scotland, with a view of comparing certain results
of his theory of the earth with actual observation.

John Playfair, 1805

THE FELLOWS OF THE Royal Society of Edinburgh left their
offices, studies, laboratories, and court chambers in the early
afternoon of March 7, 1785, and started walking toward the university library for that day's meeting. Adam Smith closed his
office at the Custom House early and walked down High Street
with his trademark cane, turning right at Nicholson Street on his
way to the university. Professors Playfair, Cullen, Robison, and
Robertson were already at the university, so they had the briefest
of strolls across the courtyard. There was great anticipation for
this gathering because this was the day of the long-awaited formal presentation of James Hutton's theory of the earth. Almost

everyone attending the talk knew Hutton personally, and they were aware of his decades-long fascination with minerals and the surface of the earth. But almost no one knew what kind of general theory this fascination had led to. They were in for quite a surprise.

No record remains of precisely who sat in the library that day to hear the first of two scheduled lectures. There were over 150 members of the society, but many—such as James Watt and Benjamin Franklin—did not live in Edinburgh. The society was also divided into two groups, the Physical Class and the Literary Class. The latter convened on a different day and may have been less motivated to listen to a talk on a topic outside its main interests. For the members of the Physical Class, though, this was going to be a riveting session, and everyone who was not ill or out of town would have found a way to attend. So at least fifty of the most learned men in Scotland assembled to hear the "famous fossil philosopher" address them. The person Hutton was probably most eager to impress, however, was Professor John Walker.

Walker had been professor of natural history at the university since 1779, and was instrumental in the founding of the Royal Society. As secretary of the Physical Class, he must have been involved in soliciting Hutton to give his talks. Walker was born in Edinburgh in 1731; like Hutton, he came from a comfortable family. He attended the University of Edinburgh in the 1740s, where he developed an interest in chemistry. Afterward, he became enthusiastic about mineralogy. In 1764, he toured the Highlands (Hutton's Highland tour was the same year) as a mineralogist assessing forfeited estates, just as Hutton did with George Clerk-Maxwell. Unlike Hutton, though, Walker was a Christian and became a minister in the Church of Scotland in the 1750s. As a young man, he had cultivated ties to prominent citizens of Edinburgh, especially William Cullen, and these connections eventually helped him to gain his appointment to the university.

As a professor at the university and Keeper of the Natural History Museum (which had a large mineral collection), Walker introduced the first known geology course in the English-speaking world, which started in 1781. Copies of his course notes make it clear that he believed in the biblical age of the earth, and that he generally followed the teachings of the most influential geologist of the day, Abraham Gottlob Werner. Because Werner's theory of the history of the earth was so well-known and accepted, it is important to describe it in some detail. The specter of Werner would haunt Hutton and his supporters for years.

ABRAHAM GOTTLOB WERNER (1749-1817) was twenty-three years younger than Hutton, and by 1785, he was known throughout Europe. Werner had spent his youth surrounded by rocks and ores—his father was the inspector of the Duke of Solm's ironworks—and had emerged as a fine mineralogist at a young age. In 1774, he published what was essentially the first rigorous field guide for identifying and analyzing minerals. On the strength of this book, Werner was given a professorship at the Freiburg School of Mines (in the German province of Saxony) in

1775, at the age of twenty-six. He was a gifted lecturer, and, much like Joseph Black's, his international fame rested on a few publications, a renowned course, and the small army of inspired students who had heard his lectures and then spread the Wernerian vision as they scattered throughout the universities of Europe. He started teaching his historical geology class in 1779, making it the first of its kind in Europe. Many of Werner's students drafted informal manuscripts based on lecture notes, which were then circulated among the international mineralogy community. By the time Hutton delivered his lectures to the Royal Society of Edinburgh, Werner's view of the history of the earth was accepted in most scholarly circles.

The Wernerian model was a synthesis of several precursors. He started with Buffon's theory of the earth, found in his thirty-four-volume *Histoire Naturelle* (1749). As described earlier, Buffon's theory was complex, but in essence argued that the earth formed when a comet collided with the Sun and the ejecta re-formed into a planet. Buffon believed that the earth had once been as hot as the Sun, but that it was now slowly cooling off and would someday stop supporting life. As it cooled, according to Buffon, the earth became covered by a universal ocean, and the features of the continents were formed as the ocean receded and evaporated. This meant that all the rocks on the earth had precipitated out of the universal ocean.

Werner next looked to Italian mineralogist Giovanni Arduino, who in 1759 published a classification scheme for all visible rock groups. "Primary" rocks were formed at Creation and had no fossils in them; they were still visible among the highest mountains, such as the Alps. "Secondary" rocks were the earliest stratified rocks and they were represented by limestone and shale; "Tertiary" rocks were fossil-filled clays and sandstones. The youngest rocks in the world were volcanic rocks, from recent eruptions. Arduino was probably influenced by the German mineralogist Johann Lehmann, who had first presented this classification but had explicitly correlated the categories with the Bible; thus, Primary rocks were formed at Creation, Secondary during the Deluge, and Tertiary after the Deluge. Arduino's scheme was less biblically oriented.

Werner embraced the idea of the universal ocean coupled with the distinctive rock groups of Lehmann and Arduino. Werner's universal ocean had been slowly receding since its inception. As time went on, different types of rocks were revealed. Primary rocks were the oldest; they were the earth's highest and were found in mountain ranges. These rocks contained no fossils because they were formed before living organisms appeared. The next group of rocks was called Transition (a group that Werner inserted into Arduino's scheme), and they were the oldest and most distorted stratified rocks, made of eroded Primary rocks and primitive dead organisms. They were often vertical and broken because they had formed on the surface of the chaotically shaped Primary rocks, or because ancient caverns had collapsed under the weight of these early sediments. The universal ocean was stormy during this stage, a condition that contributed to the extreme shapes of the layers and the many dead organisms now manifested as fossils. Secondary rocks were more recent stratified rocks, formed from eroded Primary and

Transition rocks and normally found in horizontal beds. The final type of rock was Alluvial, formed by such recent events as volcanoes or floods. Of particular significance, Werner argued that granite, a very common rock in mountain ranges, was a Primary rock, the first to precipitate from the universal ocean and therefore the oldest type of rock. Werner's theory was embraced by most of the scientific community because it seemed to explain all the rock formations found around the world. The Christian community accepted it, too, because Werner did not openly dispute the biblical time frame (although privately he had his doubts), and the universal ocean could be interpreted either as Noah's Flood or the original waters of Creation.

WHEN THE SCHOLARS STARTED arriving at the library, they were surprised to see Joseph Black, not James Hutton, sitting near the front of the room. Whether it was because of nerves or a genuine illness that afflicted him at an inopportune time, Hutton was so sick on this important day that he could not even deliver his own lecture. It is possible that he was not even in the room. Luckily, the rules of the Royal Society had forced Hutton to draft the lecture so that it could be published; thus, it was available to be read by someone else if the need arose. In a way, Hutton's incapacitation worked in his favor. Joseph Black, his best friend, agreed to deliver the talk in his place. Black was reputed to have a marvelous baritone voice and a cadence that enticed listeners, and his years of classroom experience made him relaxed and polished in front of an audience. More important, he was a lionized scientist and his agreeing to read the lecture gave it his implicit approval.

Once the fellows and guests had found their seats, Black took his place behind the lectern and began. The title was intriguing: "Concerning the System of the Earth, Its Duration, and Stability." Black's opening line immediately commanded the audience's attention: "The purpose of this Dissertation is to form some estimate with regard to the time the globe of this earth has existed, as a world maintaining plants and animals; to reason with regard to the changes which the earth has undergone; and to see how far an end or termination to this system of things may be perceived, from the consideration of that which has already come to pass." On that first day, Black described Hutton's observation that most of the land on which people now live is made up of the waste of past land, that is, stratified rocks. We know that two things have happened—"collections of loose or incoherent materials" have been consolidated, and those "consolidated masses" have been somehow elevated above the sea to form new land. Black continued by saying that the present inquiry was aimed at learning how these two related processes occurred. The rest of the first lecture focused on the formation of strata, Hutton/Black in the end discounting aqueous causes and instead proposing that heat and pressure were the reasons for consolidation.

The second lecture was delivered exactly four weeks later, on April 4. This time, James Hutton was well enough to do his own talking. Having deduced how stratified rocks formed, Hutton's next inquiry concerned the elevation of the new strata from below the seas to form new land. Once again, Hutton called on the power of heat—subterranean heat—as the causal force. Simply put, the elevation could not be the result of receding water; if it were, all stratified rocks would be horizontal, just as they had formed on the floor of lakes, seas, and oceans. Instead, it was well known that many strata were found in every degree of "fracture, flexure, and contortion"; therefore some force was pushing the strata upwards. The only available force was hot liquid rock, created by the same heat that caused the consolidation of stratified rocks. His proof of this phenomenon was "mineral veins, those great fissures of the earth, which contain matter perfectly foreign to the strata they traverse." Because these veins clearly came from below the strata, hot liquid rock must have been pushing from beneath, lifting the new stratified rocks above the sea. Hutton argued that we knew the earth had experienced this cycle of regeneration in the past because we could see fossils of "every manner of vegetable production . . . in the strata of our earth." This meant that dry land containing plants had eroded to form underwater sediments, and those sediments had later been raised as new land, where people could later find the evidence of the long-dead organisms in fossils dug from the new dry land.

At last, Hutton arrived at his remarkable conclusion, one that was based on simple logic and observation. Thus, "a question naturally occurs with regard to time; what has been the space of time necessary for accomplishing this great work?" He urged his listeners to reflect on erosion, of which they were all aware: "As there is not in human observation proper means for measuring the waste of land upon the globe, it is hence inferred, that we cannot estimate the duration of what we see at present, nor calculate the period at which it had begun; so that, with respect to human observation, this world has neither a beginning nor an end." It is important to note that Hutton was arguing that the earth is unknowably old, not eternal; the phrase "with respect to human observation" is critical in this context.

Like Charles Darwin, whose best friends did not know the full extent of his views until his paper on natural selection was first presented, Hutton had dropped a bombshell. By 1785, many theories of the earth had been put forth, so Hutton's was just one more. But nearly all previous theories had worked within the biblically prescribed 6,000 years, or they had sidestepped the issue entirely. Only one noted scholar had been as bold as Hutton. A few years before Hutton's talk, Buffon had revised his famous 1749 book, arguing that the age of the earth was 75,000 years (he arrived at this number by estimating how quickly the earth had cooled from its original molten state). However, because many features of Buffon's original hypothesis were no longer accepted, and Werner's theory was gaining popularity, the 1778 revision did not have nearly the same impact as the original. But even if it had, Buffon's 75,000 years were a simple multiple of 6,000. The earth was still very young. Hutton's idea, on the other hand, was revolutionary; simple observation of the land forced one to acknowledge that the earth was profoundly ancient, so old that one could not even hazard a guess as to its age.

Joseph Black had discovered carbon dioxide and the nature of the atmosphere; Adam Smith had properly analyzed how economies work; William Cullen had devised many important medical procedures and practical chemical discoveries for the textile industry; John Playfair was working on a text about Euclid's geometry. These were important contributions. But if

Hutton was right, his theory made the others pale before it. His would challenge the very place of humans in the cosmos.

No ACCOUNT EXISTS OF HOW Hutton's lectures were received in the early months of 1785. Playfair states that "the truth is, that [the theory] drew their attention very slowly, so that several years elapsed before any one showed himself publicly concerned about it, either as an enemy or a friend." Playfair was probably referring to written reaction because it was not until 1788 that the first published reviews appeared. The only other reference to the lectures before 1788 comes from Adam Ferguson, who in a 1787 letter to a famous French geologist, Horace de Saussure, wrote, "His (Hutton's) ideas are magnificent and, what is more precious and more difficult in science, formed with a scrupulous regard for reality."

Still, there must have been some reaction from the members of the Royal Society who sat in the library those two Fridays. Walker surely voiced his objections, since he disagreed with so many of Hutton's assertions. All that is known for sure, though, is that immediately after giving the lectures, Hutton set out on a series of field trips designed to provide evidence for his positions. Just as a teacher needs to give a lecture at least once to learn what the students do and do not grasp, Hutton learned by the audience's reaction which parts of his theory "worked" and which ones cried out for proof.

Hutton must have realized that his critics did not much care about the claim that the mineralizing process was caused by heat. However, they did care about his contention that the earth behaved in a cyclical fashion. The only way such a cycle could exist was if subterranean heat had caused hot rocks to push upward toward the surface. Thus, the major controversy was over Hutton's argument that heat was the engine that caused submerged stratified rocks to rise above the sea. Hutton's proof was that mineral veins could be found permeating stratified rocks: "In many places those consolidated strata had been broken and invaded by huge masses of fluid matter similar to lava, but, for the most part, perfectly distinguishable from it."

Hutton hypothesized that the "fluid matter" intrusions were made of granite, and that they were formed when fresh hot magma from within the earth, "the mineral region," welled up and forced its way into older stratified rocks. They then raised the strata above the sea. Werner and his followers thought exactly the opposite: that those granite veins were Primary rocks, formed at the earliest stage of the earth's history and precipitated from the universal ocean, and that the stratified rocks surrounding these Primary rocks had formed around them at some later stage. They had their unusual shape because of ancient movements that were now impossible to re-create.

JAMES HUTTON RESOLVED TO GO into the field and find proof that granite intrusions formed from within the earth and were therefore younger than the overlaying strata. He enrolled one of his friends, Sir John Clerk of Eldin, the younger brother of George Clerk-Maxwell and himself a talented mineralogist and artist, to accompany him in an effort to find strata cut by granite. They were looking for an exposure in which the shape and direction of the veins demonstrated that it came from below. Hutton knew from his 1764 trip to the Highlands with George Clerk-Maxwell that what he called "veined granite" was common there. In a later paper describing his search for granite, Hutton stated that "this question could only be determined by the examination of that species of granite upon the spot, of where it is to be found in immediate connection with those bodies which are evidently stratified." This was likely a line directed at John Walker, who stressed the need to see rocks in the field, not in the lab.

Hutton had become acquainted with the Duke of Athol, whose estate was in the town of Blair in the eastern Highlands north of Dundee and west of Aberdeen. The estate was just south of the Grampian Mountains in the part of the Highlands where the granite of the mountains often mixed with the stratified rock. Hutton remembered that the area contained many streams and rivers, which would create vivid exposures. Hutton mentioned his goal to the duke, who then invited Hutton and Clerk to visit during hunting season, when his party would be traveling to Glen Tilt. As Playfair described it, "The Tilt is, according to the seasons, a small river, or an impetuous torrent, which runs through a glen of the same name."

Playfair went on to relate what happened next:

> When they reached the forest lodge, about seven miles up the valley, Dr. Hutton already found himself in the midst of the objects which he wished to examine. In the bed of the river, many veins of red granite, (no less indeed than six large veins in the course of a mile), were seen traversing the black mica-

ceous schistus, and producing, by contrast of color, an effect that might be striking even to an unskillful observer. The sight of objects which verified at once so many important conclusions in his system, filled him with delight; and his feelings, on such occasions, were always strongly expressed, the guides who accompanied him were convinced that it must be nothing less than the discovery of a vein of silver or gold, that could call forth such strong marks of joy and exultation.

The formation "most clearly demonstrates the violence with which the granitic veins were injected among the schistus." Hutton later commented about seeing the Glen Tilt veins for the first time: "I here had every satisfaction that it was possible to desire."

The next summer, Hutton and Clerk again headed into the field. Hutton was now sixty years old and Clerk fifty-eight. This time they went to Galloway, a region at the extreme southwest of Scotland, and found more evidence of granite invading strata from below. In Hutton's words:

We therefore left the chaise . . . while we ran with some impatience along the bottom of the sandy bay to the rocky shore which was washed by the sea. . . . But breaking through the bushes and briars, and climbing up the rocky bank . . . we saw something that was much more satisfactory. . . . For here we found the granite interjected among the strata, in descending among them like a mineral vein, and *terminating in a thread where it could penetrate no farther* . . . [this] will convince the most skeptical with regard to this doctrine of the transfusion of granite.

The following summer, this time assisted by Clerk's son, Hutton visited the island of Arran, which lies in the Firth of Clyde, southwest of Glasgow. Once again, he was looking for granite meeting with stratified rock. And again, he was success- ful: "Having once got hold of the clue or caught the scent, we traced back (with more animation than could have been expected from such an innocent chase) the object of our investi- gation all the way to the Cataract rock. Great veins of granite may be seen traversing the schistus, and ramifying in all directions. I procured a specimen, which I have had conveyed to Edinburgh, though weighing above 600 pounds." Indeed, Hutton did ship a huge boulder from Arran back to his house.

The trips to Glen Tilt, Galloway, and Arran supplied ample proof of Hutton's contention that granite was formed by subter- ranean heat, that it often flowed underground, and that it forced its way into overlaying strata. More important, it showed that subterranean heat could indeed be a force that raised the land. This finding went a long way toward buttressing his theory. However, the other revolutionary idea in his scheme, that the earth recycles itself, was not necessarily proved by the granite dis- coveries. It was critical to find evidence that a cycle had actually occurred. That was what the breathtaking discovery at Siccar Point in 1788 provided. After that successful excursion along the North Sea coast, Hutton at last felt that his theory was secure.

In 1788, THE FULL-LENGTH PAPER based on the 1785 lectures finally appeared in print. The ninety-five-page document was published in the first volume of the *Proceedings of the Royal Society*. Though published three years—and four productive field trips—after the original lectures, the printed version did not mention the discoveries at Glen Tilt and Siccar Point. Hutton must have finished the manuscript in 1785, and the printers for the Royal Society of Edinburgh must have been extraordinarily slow.

Three reviews of Hutton's paper appeared in 1788, just months after the publication of the Royal Society's volume. The first appeared in the widely read *Monthly Review,* a journal that culled news and findings from recent publications. In two pages, the anonymous article summarized Hutton's argument, but then dismissively pointed out that Hutton argued for "a regular succession of Earth from all eternity! And that the succession will be repeated for ever!!" Because Hutton had not ventured a guess about the actual age of the earth, confusion was created among many of his reviewers, for they thought that he was arguing for an eternal earth, which he was not. In the previous articles the editors of the *Review* had stated their agreement with the "allegiance between Nature and Revelation, which the wisest men of all ages have discerned and admired, and which the minute philosophers of the present times have made many important efforts to destroy."

One paragraph was all it took for the *Analytical Review* to dismiss Hutton as simply another grand theorist with few facts to support his "philosophical romances."

But the *Critical Review* gave Hutton's paper a four-page critique that accurately described Hutton's positions. It did not embrace the theory, but it did not reject it, either, and it recognized the rigorous scholarship found in it.

In 1789, a new book appeared: *The Natural History of the Mineral Kingdom* by John Williams. It included a forty-page addition that was written and printed quickly to respond to Hutton. The inserted chapter summarized Hutton's arguments, refuting each in turn, and concluded with a comment about Hutton's proposed eternal earth (which he had not claimed): "The wild and unnatural notion of the eternity of the world leads first to skepticism, and at last to downright infidelity and atheism."

These essentially negative reviews seemed to have had very little effect on Hutton, who was confident in his theory, and secure with the knowledge that Black, Playfair, and Hall—and no doubt many others in the Royal Society—supported him.

However, one review did goad him into action. The hurtful critique was by Richard Kirwan (1733–1812), who published a thirty-page paper on Hutton's theory in the *Transactions of the Royal Irish Academy* (1793). Kirwan was a respected scientist who as a young man had trained to be a Jesuit priest. In 1787, after living in London for ten years, he returned to Ireland and helped found the Royal Irish Academy. At one point in his review he attacked Hutton for proposing cycles, which were "contrary to reason and the tenor of the mosaic (Book of Moses) history." He also essentially accused Hutton of being an atheist and blasphemer.

In the summer of 1793, the same year that Kirwan's paper was published, James Hutton suffered a serious illness. He was retaining urine, which could have been caused by kidney failure. He was so ill that he needed surgery, which in the late eighteenth century was anything but routine. Though very weak from the illness and surgery, Hutton resolved to write a major expansion of the 1788 paper the "very day after Mr. Kirwan's paper was put into his hands." Unfortunately, Hutton never really recovered his strength, and he suffered terribly while working on his book. This no doubt helps to explain why *The Theory of the Earth*, published in 1795, has always been regarded as poorly written (though the 1788 paper and the 1785 abstract of the Royal Society of Edinburgh lectures are both accessible and clear).

After the book came out, the sixty-nine-year-old Hutton again fell seriously ill from the same ailment. He was confined to his house, and during the winter of 1796–1797 "he became gradually weaker, was extremely emaciated, and suffered much pain." Playfair describes his last day:

> On Saturday, the 26th of March he suffered a good deal of pain; but nevertheless, employed himself in writing, and particularly in noting down his remarks on some attempts which were then making towards a new mineralogical nomenclature. In the evening he was seized with a shivering, and his uneasiness continuing to increase, he sent of his friend Mr. Russel, who attended him as his surgeon. Before he could possibly arrive, all medical assistance was in vain: Dr. Hutton had just strength left to stretch out his hand to him, and immediately expired.

Hutton's Boswell's

At any other time, the force and elegance of Playfair's style must have
insured popularity to the Huttonian doctrines; but, by a singular
coincidence, neptunianism and orthodoxy were now associated in the same
creed; the tide of prejudice ran so strong, that the majority were carried
away into the chaotic fluid, and other cosmological inventions of Werner.

Charles Lyell, 1830

JAMES HUTTON DIED ON THE evening of Saturday, March 26,
1797, at the age of seventy. One sister, Isabella, survived him. He
was buried in Greyfriar's Cemetery, the largest cemetery in the
city, lying below the south side of imposing Edinburgh Castle.
Although he had been ill for over four years, the doctor had
inexplicably made no effort to get his affairs in order. Joseph
Black, the very picture of organization and preparation himself,
was pressed into service to deal with Hutton's considerable
estate. Several weeks after Hutton's death, Black and the rest of

his still-saddened friends were shocked by an unexpected development: Hutton's illegitimate son, also named James Hutton, arrived in Edinburgh to announce his existence. Even Black, who knew Hutton better than anyone, was stunned. The second James Hutton, now about fifty years old, had lived in London all his life; he was married and had five grown children. There is evidence that Hutton had occasionally sent his son money, but no indication that he had ever met his grandchildren, and he did not provide for any of them upon his death. It is a testament to Hutton's friends, if not to him, that Black, Playfair, and others looked after Hutton's grandchildren after his death. In fact, one of the grandchildren was a motivated student, and Playfair saw to it that he was admitted into the University of Edinburgh.

Though Hutton's personal affairs were in disarray, the same could not be said of his intellectual pursuits. By 1797, he had published two of three planned volumes of the *Theory of the Earth*. The unpublished third volume would have been welcomed, for the draft manuscript contained passages about his discoveries at Glen Tilt, Arran, and Siccar Point. But it was hardly necessary. The doctor had completed all the needed fieldwork, and Clerk, Playfair, and Hall were his witnesses. He had published his theory three times: the Abstract in 1785, the long paper in the Transactions of the Royal Society of Edinburgh in 1788, and the two volumes of *Theory of the Earth* in 1795. And he had also published a short piece on granite in 1790. He no doubt died feeling that his theory and legacy were secure.

The critics were still hounding him, of course. Yet, he had dispatched Richard Kirwan at length in the 1795 book. And though an old nemesis, Jean André De Luc (1727–1817), reappeared to attack Hutton's two volumes in a series of articles published in 1796 and 1797, he seemed merely to parrot Kirwan. De Luc was from Geneva, but had lived in London since the early 1770s. He was a respected member of the Royal Society, and his aim, like Kirwan's, was to reconcile Genesis and geology. In his critique, he charged Hutton with crediting the Deity for the outlandish processes that "he has himself devised against the Mosaic account of the earth." De Luc's own theory was related to Werner's. He imagined that there had been six major epochs in the earth's history, correlating to the six days of creation, and that huge cavities in the earth had caused massive collapses, which in turn had led to dramatic recessions of the universal ocean and odd formations on the surface of the earth.

Kirwan and De Luc were respected geologists, and though their attacks were spirited, they were also old-fashioned and steeped in religion; they did not seriously engage with Hutton's rigorous scientific assertions. Had De Luc been the last to take on Hutton, his friends could have rested easy. But a new foe appeared right before Hutton died—the most talented and determined of them all. He *would* engage with Hutton's scientific assertions. To add insult to injury, this new enemy lived right in Hutton's backyard—and he was a mere child. Robert Jameson was yet another Edinburgh native and student from the University of Edinburgh who, at twenty-two years old, presented a paper at the Royal Medical Society (in Edinburgh) titled "Is the Huttonian Theory of the Earth Consistent with Fact?" As the title makes clear, the 1796 article was a direct attack on Hutton's theory. Where and why the young Jameson developed his anti-Hutton sentiment are not documented. However, it is probable that it started at the university, in John Walker's classroom. As a student, Jameson had become the protégé of the increasingly infirm Walker, who gave him responsibility for the day-to-day running of the Museum of Natural History while Jameson was still an undergraduate. A convert to the Wernerian viewpoint, Jameson was determined to prove Hutton wrong once and for all.

Unlike his mentor, who never seemed to finish the books he was supposedly writing, Jameson was prolific. Following the 1796 paper, he traveled to Ireland to meet Richard Kirwan. Still only twenty-four years old, he published his first book, *An Outline of the Mineralogy of the Shetland Islands, and of the Island of Arran,* in 1798. Arran was one of the areas that Hutton had visited during his granite-finding trips of 1785–1787, and Jameson deliberately chose to visit it to provide a different interpretation. Two years later, he published a prodigious two-volume book titled *Mineralogy of Scotland.* The book contained impressive scholarship, and it applied the Wernerian viewpoint to the geology of Scotland.

After writing two books and one article, all aimed at disproving Hutton and reinstating Werner and Walker, Jameson decided to travel to Freiberg to study at the feet of the master himself. He left for Saxony in September 1800, and studied with Werner for one year. He returned to Edinburgh more committed than ever; and now that Walker's health was in tatters, Jameson was named his assistant. When Walker died in 1803, the world's most accomplished pro-Wernerian and anti-Huttonian geologist became a formal colleague of Playfair's and the next geology professor at the University of Edinburgh.

John Playfair watched these developments with alarm. As early as 1796, the year Jameson's first paper was published, he recognized that Jameson was a potential threat to all that Hutton had worked so hard to accomplish. Moreover, Playfair was not convinced that Hutton's book had adequately confronted even Kirwan and De Luc. As evidence of how much work was left to be done, the already influential *Encyclopedia Britannica,* in its third edition of 1797, devoted twelve double-column pages to challenging Hutton's 1788 paper: "Thus we have seen, that, contrary to our author's hypothesis, the world has undoubtedly had a beginning; that our dry land has not, for ages, been the bottom of the sea; that we may reasonably suppose the deluge to have been the cause of all or most of the fossil appearance of shell, bones, & c., we meet with . . . "

Who would lead the counterattack? Joseph Black had neither the strength nor energy. John Clerk was the same age as Black and also in declining health. James Watt was busy making history in his own right, and had not stayed close to Hutton in his later years. In the end, Playfair knew that he and Hall would need to safeguard their friend's work. It appears that both men were motivated to play these roles because of the importance of the scientific questions, which they recognized as among the most intriguing in the waning years of the eighteenth century. At the same time, their affection for Hutton was unmistakable. Soon after Hutton's death, Playfair visited Arran, and he wrote to Hall from there: "The junctions I saw were I believe all visited by Dr. H. At one of them I could see the marks of his hammer, (or at least I thought so), and could not without emotion think of the enthusiasm with which he must have viewed it. I was never more sensible of the truth of what I remember you said one day when we were looking at the Dykes in the water at Leith since the Dr.'s death, 'that these phenomena had now lost half their value.'"

As a first step, Playfair decided to write a memoir of Hutton for the Royal Society of Edinburgh, similar to one he had written for Matthew Stewart ten years earlier (Stewart was the mathematician who assumed Maclaurin's chair after his death; he died in 1785, when Playfair took *his* position). While sorting through Hutton's papers, Playfair realized how much material had been committed to manuscript but never published. Also, shortly after he began the memoir, Jameson's book about Arran appeared. That book, combined with Playfair's fear that Hutton's 1795 *Theory of the Earth*—written while Hutton was tormented by intense pain—was not as strong as it could have been, caused Playfair to take an unprecedented step. He decided that he would write a new book about Hutton's theory. Thus, he embarked on *Illustrations of the Huttonian Theory of the Earth*. Other scholars have extended a predecessor's works before and since, but usually the new effort is presented as an original work. The title alone demonstrates that Playfair's goal was remarkably honorary; he was simply clarifying what his friend had already proved. In the preface to his book, Playfair wrote, with his usual modesty,

Having been instructed by Dr. Hutton himself in his theory of the earth; having lived in intimate friendship with that excellent man for several years, and almost in the daily habit of discussing the questions here treated of; I have had the opportunity of understanding his views and becoming acquainted with his peculiarities, whether of expression or of thought. In the other qualifications necessary for the illustration of a system so extensive and various, I am abundantly sensible of my deficiency, and shall therefore, with great deference, and considerable anxiety, wait the decision from which there is no appeal.

Playfair's book was published in 1802. Over 500 pages, but with a relatively large typeface and small page size, the book did not feel long, especially when compared to Hutton's lengthy two volumes. The layout itself was clever. The first part, less than 150 pages, was a distillation of the Huttonian theory and focused on stratified rocks, their consolidation and position on the surface of the earth, intrusive igneous veins, granite, the system of stratified rocks and igneous rocks, and what it all meant for the renovation of the earth. The next 400 pages consisted of separate chapters organized under the heading "Notes and Additions." A representative chapter was titled "Origin of Coal"; another was "Rivers and Lakes." These chapters demonstrated just how well James Hutton had taught geology to the mathematician; they also showed how widely read in geology Playfair had become. The chapters were filled with specific examples from all over the world, most coming from other writers, although Playfair had been doing geologic fieldwork around Scotland in preparation for writing the book (he mainly revisited the major locales of Hutton's 1785–1788 excursions). The last part of *Illustrations* addressed the various criticisms that had surfaced over the years. Playfair never mentioned Jameson by name, but he did explicitly direct comments to Werner, Kirwan, and De Luc.

Illustrations of the Huttonian System was successful. It was published in London and Edinburgh simultaneously, and it certainly attracted a wider readership than Hutton's final work (which went through only one 500-copy printing). Playfair's nephew later wrote (in 1822), "With what success [*Illustrations*] was attended we may judge from the fame and credit which have been attained by the theory, which, but for its commentary, seemed likely to be known only through the erroneous statement of its opponents."

JAMES HALL PERFORMED A function just as important as Playfair's. Seeing Siccar Point with the doctor in 1788 had been a key turning point for Hall. From then on, he was a disciple of Hutton's and he wanted to help confirm his old friend's theory. The young aristocrat was a talented chemist, and in the early 1790s he asked Hutton whether he could perform some experiments on basalt. For reasons not altogether clear, Hutton thought that any such experiment would be fruitless and urged Hall not to bother. Hall did not agree, but he did not want to openly disobey Hutton, so he waited.

Following Hutton's death, Hall saw no reason to wait any longer. The existence of subterranean heat remained a controversial topic, and Hall believed that he had a way to prove its existence through chemistry. Jameson, Kirwan, De Luc, and Werner himself all claimed that basalt—what we now know to be an igneous rock like granite—was a Primary rock, and therefore among the oldest on the planet. Hutton was convinced that it was fresh molten rock injected into older strata. The Wernerians proved their contention by pointing out that basalt, when heated and cooled in experiments, turned to glass, not the crystalline rocks that appeared in nature. Therefore, basalt must be a precipitate, formed in the universal ocean. It is important to remember that the followers of Werner merely had to prove that the upstart Hutton was wrong because he was confronting their established theory.

Hall had an ingenious idea: What if the heated basalt was cooled only very slowly, as would probably happen within the earth, where Hutton reasoned that basalt formed? Hall collected fifteen separate samples of basalt from Scotland, England, and the Continent, heated them to very high temperatures, and then let them cool slowly. Sure enough, the basalts re-formed as crystals, not glass. With Hall's connections in Paris and at the Royal Society in London, word of the experiments spread through the geology community and were quickly recognized as significant. However, the resulting crystals still did not look like their parent rocks (the basalts needed to cool much more slowly than Hall had allowed). Therefore, though this experiment proved the Wernerians were wrong in believing that basalts always cooled into glass, it did not conclusively prove Hutton correct.

But Hall's next experiment would do just that. It would take about six years to complete and it is now considered the beginning of experimental geology. Hall's goal was to prove that limestone would not disintegrate if heated under great pressure. Recall that Joseph Black had discovered carbon dioxide by heating types of limestone; his specimens lost nearly half their weight when heated, and he was then able to prove that what was lost was carbon dioxide. So this was well known. And it was precisely what critics pointed to when Hutton argued that subterranean heat caused sediments to consolidate: If limestone, one of the most common of all stratified rocks, became completely transformed when heated, how could heat be the "mineralizing" factor? Hutton had argued that pressure kept the components together and prevented disintegration, but this had not been proved.

Hall performed over 500 experiments designed to prove that limestone did *not* disintegrate when heated, as it does on earth, as long as there was enough pressure to hold it together. He had to design every facet of the experiment because nothing like it had ever been attempted. In addition, he had to assemble his own instruments and equipment, which included a special high-temperature thermometer, and gun barrels (at the time, the only objects built to withstand high pressure). Hall placed the rocks in the gun barrels, heated them to tremendously high temperatures, and then measured the loss of weight. In his most successful experiments, with the heat as high as 1,000 degrees Celsius and pressure measurements equivalent to a column of salt water 2,700 meters (almost 2 miles) high, the minerals lost essentially no weight. This was a triumphant result.

Hall's new success was again widely publicized. His reputation as a careful chemist meant that the results were taken seriously. Unfortunately, because of their complexity, the experiments were nearly impossible for others to replicate, so they could not be independently confirmed. Until similar results were achieved by others, and they finally were decades later, Hall's findings could not be considered unassailable. Clearly, there was more work to do.

PLAYFAIR'S *Illustrations of the Huttonian Theory of the Earth* was published in 1802, and Hall's limestone experiments were completed in 1804–1805. These two men had done more than anyone could ask to protect another's legacy. And for most of the first decade of the nineteenth century, it looked as though the Huttonians had won the battle against the Wernerians. But Werner's advocates were simply too numerous and determined for Playfair and Hall to withstand for long. Werner lived until 1817, teaching his course regularly and creating new apostles each time. Robert Jameson was probably even more instrumental in keeping Werner's ideas predominant. He was the only professor at the University of Edinburgh teaching the geology course, and it was the largest course of its type in the world. Records of the period show that he taught between 50 and 100 students each year (Werner taught only about 20 annually), creating a small army of Wernerians by the end of the term. Rubbing salt into the wound, Jameson did more than espouse the tenets of Werner's system; he also took pains to criticize Hutton's system at every turn. In addition to his teaching, Jameson published his most successful book yet in 1808. Titled *Elements of Geognosy,* this technical book was the most scholarly and refined presentation of Abraham Werner's system.

Although the tension in Edinburgh was palpable, the geological community in England officially turned its collective back on the controversy. The French Revolution had created a tremendous backlash of conservatism in England that started in the early 1790s and continued for several decades. The conservatism spread in all directions, infecting even the sciences. Kirwan and De Luc's critiques of Hutton were emblematic of this backlash. Then, in 1807, the prominent geologists of England founded the Geological Society of London. The founders of this body were weary of the Hutton/Werner debate and essentially told their members that the field needed facts and observations, not theories. The purpose of their society would be to provide the needed facts. Therefore, the society encouraged and supported specific and detailed investigations of geological formations.

In 1808, the same year that Jameson's book appeared, a development on the Continent complicated the picture. Two prominent French scientists, Georges Cuvier and Alexandre Brongniart, presented a paper that summarized the results of their close examination of the strata around Paris. They had discovered something puzzling: Nine distinct episodes had been revealed in the strata. Most significantly, the fossils found in the strata alternated between saltwater and freshwater organisms. The collaborators were followers of Werner, but this discovery shook their belief in his system. Werner's universal ocean had receded only once, so the alternation between salt and fresh water was impossible. But this had certainly occurred around Paris.

When Cuvier published his book on the joint Cuvier/ Brongniart research in 1812, he presented a new theory in which the earth had been beset by a series of catastrophic floods, at least six times in its history. These catastrophes accounted for the otherwise unexplainable strata around Paris. But the previous deluges had occurred so far in the past that they could not be analyzed or otherwise investigated. Cuvier wrote that it was not possible to "explain the more ancient revolutions of the globe by means of still existing causes. . . . The thread of operation is here broken, the march of nature is changed, and none of the agents that she now employs were sufficient for her ancient works." This view soon became known as Catastrophism. Because Cuvier argued that "the thread of operation is here broken," he did not attempt to address the question of how old the earth was.

The irrepressible Robert Jameson quickly translated Cuvier's book into English. The Jameson edition was widely read, going through five editions from 1813 to 1827. Jameson was able to overlook the problems that the book presented for strict Wernerism, the most significant being that it dramatically altered the original concept of one, and only one, universal ocean. The crucial factor, in Jameson's eyes, was that Cuvier emphasized that the final catastrophe was the source of the earth's current geology. This final deluge could be interpreted as the universal ocean. Furthermore, Cuvier's contention that the type and scale of the earth's operations in the past were different from forces currently at work was in direct opposition to the ideas Hutton and his followers had proposed.

While Cuvier's work was energizing the Wernerians, James Hall inadvertently contributed to the obscuring of Hutton's vision. One part of Hutton's theory that troubled him was its inability to explain the appearance of huge boulders (called *erratics* by geologists) in various parts of Europe. Hall came to believe that tsunamis—overwhelming tidal waves—might be the cause (Hall even placed dynamite under water to see whether he could create small-scale tsunamis for study; in the end, all he created were big splashes of water). He was particularly certain that something like a tidal wave must have crossed Scotland, from Glasgow to Edinburgh. The erratics that Hall was trying to explain had actually been transported by glaciers. They were remnants of the last ice age. Nonetheless, his 1815 article, "On the Revolutions of the Earth's Surface," appeared to some readers to question the fundamental Huttonian belief in slow and continuous action of everyday geologic processes.

The year 1815 also marked the end of the Napoleonic Wars in Europe. Toward the end of that year, John Playfair embarked on a lengthy tour of Europe in preparation for a second edition of *Illustrations,* a revision that was going to be a thorough reworking of the material. So much had occurred since 1802—the maturation of Playfair's own ideas, Hall's seminal experiments, Cuvier's important discoveries in Paris and his theory of catastrophism to explain them, and even Jameson's specific contributions—that the planned revision would likely be an entirely new book.

He spent seventeen months on the Continent and traveled all over France, Switzerland, and Italy, covering more than 4,000 miles. Back in Edinburgh in the middle of 1817, the writing of the second edition was interrupted by an invitation from the editors of the *Encyclopedia Britannica* to write a comprehensive review article on the state of the mathematical and physical sciences for one of their supplements. This was an offer the mathematician could not refuse, and he enthusiastically took up the challenge. The second edition would have to wait a few months. Sadly, the second edition never appeared. In fact, it was never even begun. Soon after completing the first draft of the article for *Britannica* in mid-1818, Playfair began to suffer from a "disease of the bladder" that nearly incapacitated him. Over time, he recovered somewhat, only to have the disease return with a vengeance in June 1819. One month later, John Playfair was dead at the age of seventy-one.

In 1822, his nephew, James Playfair, wrote a memorial, similar to the one John Playfair had written for Hutton in 1805. One telling passage shows the perilous state that Hutton's theory of the earth was in now that its leading expositor was deceased:

> It has been said that the illustration of a theory of the earth was but a profitless employment for so accurately thinking a philosopher, and that the task aught have been left to more imaginative minds, whose speculations would have afforded equal pleasure to those who delight in forming fabrics of theory on insufficient foundations. It is true that even the lucid commentary of Playfair does not establish the Huttonian as a general and undeviating theory, in an undoubted and indisputable situation.

From a close relative, this was hardly an endorsement for a body of work that so dominated the energy of John Playfair.

At the beginning of the 1820s, then, with Jameson still very active at the University of Edinburgh, Cuvier in his prime in Paris, and James Hall getting on in years, there was little hope that James Hutton's theory of the earth would ever become widely accepted. It would take another rigorous thinker—from an entirely new generation—to embrace the undeniable logic behind the ideas and recognize Hutton's true brilliance.

10

The Huttonian Revolution

> I had brought with me the first volume of Lyell's
> *Principles of Geology,* which I studied attentively; and this
> book was of the highest service to me in many ways.
>
> *Charles Darwin, 1876*

IN THE FALL OF 1824, THE ELDERLY James Hall (now sixty-three years old) hosted a young geologist by the name of Charles Lyell for several days at his estate in Dunglass. Lyell was returning to London after several weeks of research in an area north of Edinburgh. Hall so enjoyed the young man's knowledge and enthusiasm that he even took him by boat to see Siccar Point, just the way he, Hutton, and Playfair had seen it. In fact, Hall was now about the same age that Hutton had been, and Lyell the same age as Hall, on that exciting day in 1788. Siccar Point worked its magic again—just as Hall had been converted to Huttonianism years before, Lyell was poised to become a believer.

Charles Lyell (1797–1875) was destined to be connected to James Hutton. He was brought into the world only eight months after the doctor's death and within 20 miles of Glen Tilt. Though born in Scotland, Charles grew up in southern England, the oldest child of a wealthy family. He had a pleasant and comfortable childhood, in which he developed a keen interest in nature. In 1816, the eighteen-year-old Lyell prepared to enter Exeter College at Oxford University. During the summer of that year, the inquisitive youth read Robert Bakewell's *Introduction to Geology,* a popular text published in 1815. Bakewell discussed Hutton's theory and his idea that the earth was indefinitely old, but the book endorsed a theory that was a refinement of Werner's scheme; it featured the universal ocean, Primary/Transitional/Secondary rocks, and a young earth that had experienced catastrophic changes during its short history. Though he entered Oxford to study the humanities and work toward a career in law, Lyell was drawn to geology, thanks to Bakewell's book. Once at Oxford, a new inspiration took over: one of the most dynamic professors who ever taught there, William Buckland (1784–1856).

The Oxford professor was a fascinating character, and in many ways he represented the state of geology, at least in England, at the start of Lyell's career. Buckland was among the half-dozen natural philosophers who founded the Geological Society of London in 1807. That learned body was decidedly anti-theory, committed to gathering solid information about specific geological facts, formations, and processes.

William Buckland did not quite follow that creed. A very devout man, he had developed his own theory of the earth, which was related to Werner's and Cuvier's. His theory became influential because he was the first academic geologist in England, and he was the first to teach a geology course at an English university (the topic having been taught in Edinburgh since 1781). Buckland's theory was a safe synthesis of Cuvier's and Werner's work; he argued that the Deluge, which had occurred when biblical scholars said it did, about 5,000 years ago, had formed the earth that we now inhabit, and that nothing dramatic had happened since then. Thus Werner's theory of a universal ocean and the precipitation of existing land from it still held. But according to Buckland, who took his cue from Cuvier, Noah's Flood had been but the last of several huge convulsions of the earth. The previous disturbances were impossible to understand or to investigate, so it was not worth even dwelling on them. Only God knew what had occurred. Buckland would write in 1820:

Again the grand fact of a universal deluge at no very remote period is proved on grounds so decisive and incontrovertible, that, had we never heard of such an event from Scripture, or any other authority, Geology of itself must have called in the assistance of some such catastrophe, to explain the phenomena of diluvian action which are universally presented to us, and which are unintelligible without recourse to a deluge exerting its ravages at a period not more ancient than that announced in the Book of Genesis.

Lyell was well exposed to Buckland's thinking, taking three courses from him, one each spring from 1817 to 1819. Buckland recognized his student's ability as a scholar, and the two of them would remain friends for years.

After finishing at Oxford in the spring of 1819, Lyell moved to London to study law at Lincoln's Inn, but he also continued to investigate geology. He immediately became a fellow of both the Geological Society and the Linnean Society, thanks to the enthusiastic recommendation of his Oxford professor. In early 1820, severe vision problems forced Lyell to take a leave of absence from his legal studies. For a diversion, he accompanied his father on a visit to Europe that summer. This was his second trip to the Continent. During the first one, taken with his entire family in 1818, he had seen Paris; this second tour took him through the Alps and most of Italy, where he observed inspiring geology. From this point on, a series of events was going to set Lyell on a collision course with his friend and teacher Buckland.

One year after seeing the Alps and volcanoes of Italy, in October 1821, the always-inquisitive Lyell happened to visit Gidean Mantell, a surgeon who was famous in southeast England for his interest in fossils. Mantell lived at Lewes, in Sussex—in a part of England dominated by massive chalk hills, the South Downs. Beneath the chalk, which is a limestone known to be formed in a deep-sea, saltwater environment, Mantell had found something curious. He had discovered the fossils of numerous land plants, land vertebrates, and freshwater shells, all *below* the chalk. Mantell concluded that the formation he was exploring was an ancient river delta that had later been submerged by the sea and remained there long enough to be covered by layers of chalk hundreds of feet thick. Then the entire formation had been raised above the sea, and the chalk and underlying sandstone eroded away to reveal the remains of the river delta as stratified rocks. Such a phenomenon could not be explained by Werner's theory, but it could be explained by Hutton's, where just such a succession was predicted. As one Lyell scholar has commented, "By providing clear evidence of downward and upward movements of the land, the strata supported Huttonian geology rather than the Wernerian view."

Lyell started doing his own research in the area, broadening his scope to include the Isle of Wight. In 1822–1823 he confirmed for himself the same succession of events that Mantell had discovered. An admirable characteristic of Lyell's was that throughout his career he tried to see and confirm, firsthand, exposures that others had seen before him. Lyell brought Buckland with him to the Isle of Wight in early1823, and just as his confidence in past ideas was wavering, so, too, was his confidence in the abilities of his old teacher. Lyell wrote: "I should have been at a loss to conceive how so much blundering could have arisen [in Buckland's research] if I had not witnessed the hurried manner in which Buckland galloped over the ground."

The next big shake-up of Lyell's beliefs occurred in the summer of 1823. Now a practicing lawyer, Lyell traveled to Paris as a representative of the Geological Society to meet with his French counterparts. Here he met the father of catastrophism, the renowned Georges Cuvier. He also became friends with Constant Prevost, who would play a major role in his future thinking. Prevost showed him some outcrops near the city that demonstrated a process of freshwater-saltwater alternation, similar to what Mantell and Lyell had found in southern England. What made these formations even more interesting than the English ones, though, was that evidence from the fossils showed that at one point the two environments had mixed together *at the same time.* The saltwater seas that eventually overwhelmed the freshwater formation had covered the area only very gradually. Prevost believed that Cuvier's catastrophes were not needed to explain the past in this part of the Paris Basin. Sussex and the Isle of Wight had disproved Werner's universal ocean, since Werner's ocean retreated only once. Perhaps Cuvier's and Buckland's catastrophes were not needed to explain the geological changes of the past either.

Fittingly, the discovery that turned Lyell away from catastrophism once and for all occurred in Scotland, near Glen Tilt. In the summer of 1824, Lyell went back to his birthplace, Kinnordy, to do research at a marl pit there. Two small lakes, Lochs Bakie and Kinnordy, had been drained for marl, which was still used as fertilizer, just as it was in Hutton's farming days. The dry lakes revealed freshwater limestones, similar to those that Lyell had seen in the Paris basin, and they contained beautifully preserved freshwater fossils. The lakes in Scotland had only recently been drained, and they were clearly modern, geologically speaking. Cuvier had argued that the Paris freshwater limestones were formed only in the deep past, and could not occur in the modern age. Lyell had just found that Cuvier was wrong— a previously supposed ancient process had occurred recently.

After he had finished his work in Kinnordy and was on his way back to London, Lyell paid his visit to James Hall. Since first visiting with Mantell three years earlier, Lyell's belief in Buckland's synthesis of Werner and Cuvier had been shaken. Now he had met the last of the great Huttonians and seen the one-of-a-kind unconformity at Siccar Point. It is not known when Lyell first read Playfair's *Illustrations,* but now he certainly picked up the book again.

AFTER THE 1824 EXCURSION TO Scotland, Lyell returned to London, and remained there for an extended period. Over the next few years he did a great deal of writing, and his articles started to reflect his changing views. In 1826, Lyell published an important series of articles that focused on his work in southern England and Scotland. In the last of the three he stated, after discussing evidence from Chile that an earthquake had caused the land to rise significantly above sea level: "No one can reflect on the above statement without being tempted to inquire whether the causes now in action are, as Dr. Buckland has supposed, 'the last expiring efforts of those mighty disturbing forces which once operated,' or whether as Hutton thought, they would still be sufficient in a long succession of ages to reproduce analogous results."

The stage was now set for the complete transformation of Lyell from a Cuvier skeptic to a Huttonian convert. Lyell would at last be won over after learning firsthand about the power of volcanoes, a phenomenon that Hutton had avoided because of the Wernerians' preemptive claims that they were strictly shallow surface events (that is, coal or some other volatile mineral just under the surface of the earth was somehow ignited). In 1827, Lyell asked the editor of the *Quarterly Review* to let him review a new book by his friend George Scrope, titled *Memoir on the Geology of Central France.* He was drawn to it because Scrope focused on an area in south-central France that was known for its unusual geologic formations, such as cone-shaped hills and deep gorges cut by rivers. In his book, Scrope described a region that had experienced wave after wave of past volcanic activity. The formations could not be classified as either pre-Deluge or post-Deluge, as catastrophists would wish to do. The well-exposed outcrops, revealed by deeply cutting, fast-flowing rivers, showed a regular pattern of lava flows, layers of river gravel, then more lava flows, then more gravel layers, and so on. The successive layers of basaltic lava and gravel were not deformed and there was little doubt about the order in which the volcanic eruptions had occurred.

Scrope went on to present a hypothesis. He believed it was probable that volcanic activity not only raised land, but that it also caused land nearby to subside later, after the intense magma activity ceased. The seesawing earth could provide a hint for how the freshwater-saltwater alternation might have occurred in the Paris Basin.

Shortly after reading Scrope's book, Lyell set out to explore the Auvergne region of south-central France himself. He arrived there in May 1828, and was soon able to confirm all that Scrope had observed, even extending the analysis by more accurately sequencing the past volcanic activity. From central France, Lyell continued on to Italy—the country of choice for a geologist recently convinced that volcanoes held the key to a theory. Once in Italy, it did not take Lyell long to find what he was looking for. He triumphantly wrote to his sister from Naples: "I will let the world know that the whole Isle of Isk, as the natives call it, has risen from the sea 2,600 feet since the Mediterranean was peopled with the very species of shell-fish which have now the honor of living with, or being eaten by, us—our common oyster and cockle amongst the rest."

Essentially, what Lyell had found was more dramatic proof of the uplift of stratified rocks that Hutton had postulated in 1785. Hutton had needed only to see veined granite at Glen Tilt; Lyell needed to see something more dramatic—evidence that the strata had been raised half a mile in a relatively short amount of time. Regardless, Lyell left Italy shortly afterward and returned to England to continue work on the book he had started before his trip.

Truly inspired by what he had seen and learned on the Continent, Lyell finished the first volume of his book in about one year. Volume 1 of the *Principles of Geology, Being an Attempt to Explain the Former Changes of the Earth's Surface, by Reference to Causes Now in Operation* was published in London in July 1830. Two more volumes would follow, in 1832 and 1833. The book, using Hutton and Playfair as starting points, and then effectively synthesizing a huge body of work, became the defining book for the still-young field of geology. For at least the next 100 years, *Principles of Geology* would be the standard reference for students and researchers of geology. Comparable to Adam Smith's *Wealth of Nations* in its importance and immediate impact, it would go through five editions in the 1830s, and eleven editions overall during the author's lifetime. Most important, it finally, and firmly, established the earth as immeasurably ancient.

THE IMPORTANCE OF LYELL'S *Principles of Geology* was not lost on Charles Darwin, who brought a copy with him aboard the *Beagle,* the research vessel that would be his home from 1832 to 1836. He was given a copy of the first volume as a gift from one of his Cambridge University professors before the ship set sail, and he decided to read it before they reached the first destination on the itinerary. On the fourth page of the 500-plus-page book, Lyell introduced his readers to James Hutton, stating that Hutton was the first scholar to treat geology as its own subject, and the first to separate it from cosmogonies. "His doctrine on this (point) was vehemently opposed at first, and although it has gradually gained ground, and will ultimately prevail, it is yet far from being established," said Lyell. Then, after reviewing the theories and discoveries of previous thinkers in geology, the author stated that "[Hutton] was the first . . . to explain the former changes of the earth's crust, by reference exclusively to natural agents. Hutton labored to give fixed principles to geology, as Newton had succeeded in doing to astronomy." Finally, Darwin arrived at the key passage:

> If any one ventured to doubt the possibility of our being enabled to carry back our researches to the creation of the present order of things, the granitic rocks were triumphantly

appealed to. On them seemed written in legible characters, the memorable inscription Dinanzi a me non fur cose create se non eterne, and no small sensation was excited when Hutton seemed, with unhallowed hand, desirous to erase characters already regarded by many as sacred. "In the economy of the world," said the Scotch geologist, "I can find no traces of a beginning, no prospect of an end;" and the declaration was the more startling when coupled with the doctrine, that all past changes on the globe had been brought about by the slow agency of existing causes. The imagination was first fatigued and overpowered by endeavoring to conceive the immensity of time required for the annihilation of whole continents by so insensible a process.

Charles Lyell's book went on, in beautifully written detail, to prove this statement.

Darwin did not know what to think. He knew all about Hutton, but only as an object of ridicule. His first university geology professor had scathingly criticized the Huttonian viewpoint, and of course the seventeen-year-old student had believed him. And just who was this teacher? None other than Robert Jameson. For two years, from the fall of 1825 to the spring of 1827, Charles Darwin had attended the University of Edinburgh. His father, who had received his medical degree there, pushed Charles and his older brother Erasmus to study medicine at the same institution—Charles was to begin instruction and Erasmus was to finish his. Darwin detested the lecture style of teaching practiced at Edinburgh, and he would leave without a degree after two years. However, during those four semesters, he took two courses that he later wrote about: Thomas Hope's (Joseph Black's successor) chemistry course and Robert Jameson's geology course. Hope (1766–1844) had been friends with Hutton, Black, and Playfair, and he was a confirmed Huttonian. He devoted several weeks of his class to Hutton's theory of the earth, and he showed how chemistry was a tool for its proof.

However, several weeks could not counterbalance Jameson, who had an entire course with which to indoctrinate students. Jameson, although a boring lecturer, was an enthusiastic and persuasive scholar in face-to-face encounters. He urged his students to spend three hours a week in the natural history museum looking at rock specimens (which Darwin enjoyed), and he often took his students into the field around Edinburgh. Darwin left this anecdote in his autobiography, written after he had become a follower of Lyell/Hutton: "Equally striking is the fact that I, though now only sixty-seven years old, heard Professor Jameson, in a field lecture at Salisbury Crags . . . with volcanic rocks all around us . . . say that it was a fissure filled with sediment from above, adding with a sneer that there were men who maintain that it had been injected from beneath in a molten condition."

Clearly, as late as 1826, Jameson was teaching unreconstructed Wernerism, deliberately going to the very spot on Salisbury Crags where Hutton had spent so many hours speculating about his theory, simply to deride him and promote the opposite position. The future father of evolution was one of 200 students who took Jameson's course that term—its popularity had grown since the early 1800s. Darwin was even a regular guest at meetings of Jameson's Wernerian Society while at Edinburgh.

After leaving Edinburgh, Darwin was influenced by another geologist whose beliefs also differed greatly from Lyell's. Adam Sedgwick (1785–1873) was the geologist at Cambridge University, where Darwin matriculated next. Darwin never took his course, but he did serve as his assistant on a three-week field excursion to Wales during the late summer after he graduated in 1831. This fieldwork was intended to prepare him for his naturalist's duties on the *Beagle,* but he also received instruction in catastrophist thinking. Sedgwick was not the proselytizer that Jameson was; as a member of the Geological Society of London, he was more interested in fact collecting than in theorizing. However, his fundamental beliefs were similar to those of his Oxford colleague, Buckland: that the earth had experienced a succession of cataclysms, the last of which was Noah's Flood— the cause of the geology all around them.

A WEEK OR SO AFTER CHARLES DARWIN started reading the *Principles of Geology,* the *Beagle* dropped anchor at the first port on its itinerary, Porto Praya on the volcanic island of St. Jago in the Cape Verde Island chain. The extinct volcanoes that formed the islands, possessions of the Portuguese government, were located about 300 miles west of the North African coast. Though hilly, the island appeared desolate and nondescript to Darwin at first. One reason he had taken the position as the ship's naturalist was to see the lush tropical paradises he had read about as a student. It did not appear that St. Jago was going to fulfill that desire.

But it did fulfill his desire to see unusual sights. After several days of exploring the exposed black landscape of the lava-formed island, taking notes, and gathering specimens, Darwin came upon a beach, where he began collecting sponges and corals. Some distance from the beach was a low but massive hill. On its face was a distinct white band at least 30 feet above the ground. From where he stood, it looked as if a stripe had been painted across the exposed rocks. When he climbed the hill and got on his hands and knees for a close look, Darwin discovered that the band consisted of shells and coral; they were so delicately preserved that they resembled the fresh ones he had just collected. How on earth did this stratum of dead ocean life come to be raised so gently 30 feet above the level of the waves?

The explanation could not be that the water level had fallen, because though the white band was visible all along the shore, its height varied. Clearly the land had been raised up somewhat unevenly, and before it had been elevated the white band had been the beach. Jameson would have had no reasonable explanation, and Sedgwick would have called on a past catastrophe to explain it. But would not a catastrophe have destroyed the near perfection preserved in this stratum? On the other hand, Lyell's book described processes whereby land was uplifted gently. Darwin later related that finding this particular spot on St. Jago was a defining moment for him; the impressions formed there would "never be effaced." From this day on, Darwin believed in the science of Lyell—which was based on the science of James Hutton—a science that envisioned a dynamic yet ancient earth, constantly undergoing slow and subtle changes caused by the natural processes of erosion, tides, storms, volcanic eruptions, earthquakes, tidal waves, and uplift due to subterranean heat. Darwin later wrote to one of his friends back in England that he had "become a zealous disciple of Mr. Lyell's views."

From the day he found the white stratum on St. Jago, Darwin viewed the world differently. Nothing he saw during the rest of his voyage caused him to doubt his new perspective. In fact, he was "tempted to carry parts to a greater extent, even than [Lyell] does."

The *Beagle* continued its circumnavigation. Over the next four and one-half years, Darwin would see most of the Southern Hemisphere, collecting bags full of specimens—animal, insect, plant, and mineral—and sending them on to England for analysis. The ship made stops at Brazil, Patagonia, and the Falkland Islands, sailed through the Straits of Magellan, and then on to Chile. In Chile, Darwin crossed the Andes into Argentina. From the west coast of South America, the *Beagle* then journeyed to the Galapagos Islands, off the coast of Equador, and from there across the Pacific Ocean to Tahiti, New Zealand, Australia, and Cape Town. The *Beagle* docked in Plymouth in October 1836, the ship's mission finally completed.

It would soon become apparent that the most significant stop made by the *Beagle* was the Galapagos Islands. Darwin was there for over a month in the early autumn of 1835, and among the many specimens he collected were the different types of finches that he found on these tiny volcanic islands. Nine months later, the Galapagos were still on his mind; he would write in his diary, "When I see the islands in sight of each other and possessed of but a scanty stock of animals, tenanted by these birds but slightly differing in structure and filling the same place in nature, I must suspect they are varieties . . . if there is the slightest foundation for these remarks, the zoology of the archipelagoes will be well worth examining: for such facts would undermine the stability of species."

Back in England, it would not take long for those tiny birds to start a scientific revolution. Darwin thought that they were simply different varieties of the same species; but, as he did with all his collected specimens, he sent them to an expert for formal analysis. John Gould (1804–1881), who was the taxidermist for the Zoological Society, and known as one of the most careful ornithologists in England, took on the task. In March 1837, Gould told Darwin that the birds were not different varieties; *they were different species.* The importance of this finding was not lost on anyone; there was even a front-page article in the *London Times* about it. For Darwin, Gould's news was a bombshell, as significant as the day on St. Jago when he became aware of the great age of the earth and its slow but constant changes. Now it was clear that animals, perhaps all life, also experienced slow but constant changes, so much so that over time, animals that were separated from each other, as the finches on the Galapagos were, became distinct species. In July of that same year, Darwin started a new notebook with the heading "The Transmutation of Species."

The third and last flash of insight for Darwin's theory of evolution occurred on September 28, 1838 (Darwin's life is so well documented, through his careful notes and diaries, his numerous letters, and his own autobiography, that we actually have precise dates for these crucial turning points). On that day, seeking relaxation from his intense studies, which were at fever pitch in the several years after the *Beagle* returned, Darwin read a famous work that he had never read before: Thomas Mathus's forty-year-old *Essay on the Principle of Population* (1798). Darwin was having difficulty with the transmutation of species and his thinking was stalled at the time because he could not devise an engine that drove the process, much as Hutton had at first struggled to find a mechanism for the renovation of land. Darwin already knew that different varieties of organisms were born every day, and that the variations were inheritable; that is, they were passed down to offspring (biologists would not understand the specifics of mutations and genetics until well into the twentieth century). For example, in the finch population on the Galapagos, every day a newborn appeared that had a slightly longer beak than its parents, or slightly more colorful plumage, and that newborn would grow to be a reproducing adult whose own offspring would also have the longer beak or more colorful plumage.

Mathus now provided the reason why simple variation, over time, meant new species: overpopulation. Because there was not enough food, shelter, and territory for every offspring of an organism, there was competition among them for these scarce commodities; the ones who were more able to acquire these resources would be more likely to find a mate and reproduce successfully than the ones who were less able. If a variation such as a longer beak, which gives the bird the ability to find and eat seeds found deeper in crevices, helped in the quest for more resources, it would eventually be propagated through an entire population. And over great expanses of time, if the population became separated from others (as happened with the finches in the Galapagos), these small changes would lead to separate populations that would no longer be able to reproduce with one another: new species.

Charles Darwin sketched out his theory of evolution in a 30-page manuscript in 1842. Then he expanded this first effort into a 200-page essay, copies of which he gave to only a handful of his closest friends, his scientific confidants, in 1844.

Though not one of Darwin's confidants, Charles Lyell had become a close friend of the naturalist soon after the *Beagle* landed in England in 1836. Darwin's letters from the ship to his colleagues in England generated tremendous interest among the science community in London, and Lyell was the current star of that community, thanks to the impact of his book. Lyell called on Darwin soon after the *Beagle* landed, and the young man was honored to be so well regarded by a fellow scientist whose work he admired. Their friendship remained close for the rest of their lives, even though for many years Lyell could not bring himself to admit that natural selection led to the creation of new species; he finally did so in the tenth edition of *Principles of Geology* (1867–1868).

Nonetheless, Lyell played an integral role in the official debut of the theory of evolution and the publication of Darwin's *Origin of Species*. On June 18, 1858, Darwin received a letter and twenty-page paper from Alfred Russel Wallace, a young naturalist with whom Darwin had been corresponding for several years. Wallace was hoping that Darwin would read the paper and then send it on to Charles Lyell to consider for publication in the journal of the Linnean Society (Lyell was the president of that organization). To Darwin's shock and deep dismay, the enclosed paper was a sketch of a theory almost identical to Darwin's theory of evolution by natural selection. When he sent the paper to Lyell, Darwin wrote, "If Wallace had my MS sketch written out in 1842, he could not have made a better short abstract." Lyell and Darwin's friend, the botanist Joseph Hooker, acting like King Solomon, together made one of the wisest decisions in the history of science: They decided that Wallace's paper would be formally read along with one by Darwin (based on his 1844 treatise), thus guaranteeing that the two men would receive joint credit for the discovery. On July 1, 1858, extracts from Darwin's and then Wallace's papers were read to about thirty members of the Linnean Society (Darwin was not present for the reading, his last child having died of scarlet fever just three days before). Darwin's secret theory was finally out; he now shared its authorship with Wallace.

The theory of evolution by natural selection is today thought of primarily as Charles Darwin's because it is well known that he had distilled it fifteen years before Wallace. But, more significantly, Darwin followed up the paper by quickly finishing his *On the Origin of Species by Means of Natural Selection,* published in November 1859 (Darwin's publisher was John Murray, who had also published Lyell's book). Lyell acted as Darwin's literary agent, and was one of perhaps a dozen men who read the book in proof sheets. Darwin had been working on a book since 1856, but when Wallace's paper appeared, he realized that he was out of time. The book ended up much shorter than Darwin first envisioned, which proved to be a blessing; unlike many treatises of the time, it was accessible and concise, and used only the best, most relevant examples. Every page in the book was vivid and substantive.

Darwin was filled with anxiety and concern about the book's reception, but the day before it was released to the public he wrote to Lyell: "I have asked myself whether I may not have devoted my life to a fantasy. Now I look at it as morally impossible that investigators of truth, like you, . . . can be wholly wrong, and therefore I rest in peace."

Charles Darwin need not have worried, for shock waves produced by the book were immediate and lasting. The radical implications of *The Origin of Species* were soon felt in nearly every field of intellectual inquiry, and they reverberate even today. The first and arguably most important insight for Darwin on his journey of discovery was that the earth was old beyond calculation; how else would evolution have had time to work? And that is James Hutton's ultimate contribution.

EPILOGUE

T HE SUCCESS OF LYELL'S *Principles of Geology* was so pronounced that the biblical geologists and catastrophists finally threw in the towel. By the mid-1830s, Buckland had dropped his dependence on the Deluge as an explanation for the earth's geologic formations. Even Jameson softened his positions, and his later papers reflected his acceptance of the Hutton-Lyell view. The Huttonian revolution was won, and the discipline of geology, finally freed from the blinkers of catastrophes, deluges, and universal oceans, could now get on with the difficult task of determining just what had occurred over the incredible expanse of geologic time.

If the earth was not young, and if catastrophes did not mask the past, was it possible to arrive at a figure for the earth's age? James Hutton never attempted an estimate. He knew that the means for projecting an accurate date did not exist; therefore to venture a guess was irresponsible. Besides, he was critical of theorists who tried to calculate the beginning of time, and he was determined not to fall into the same trap.

Charles Lyell followed his lead and steadfastly refused to commit himself.

However, Lyell's friend Charles Darwin did venture something that was perceived as a guess, and it caused him much more trouble than it was worth. In the first edition of *The Origin of Species,* while making a point about the vast amount of time needed for even the most subtle of geologic changes, Darwin gave as an example his calculation for how long it took to form the Weald Valley in the south of England. He incautiously wrote that the valley took approximately 300 million years to form. With armies of critics biting at Darwin's heels, any assertion less than completely supported by hard facts was seized on as a weakness in Darwin's thinking. Several scholars felt that 300 million years was an absurdly high number, and they criticized Darwin for soft thinking. Because this point was not essential to his arguments, Darwin dropped the estimate by the third edition of his book.

Just a few years after the initial publication of Darwin's *Origin,* the first genuine attempt to calculate the age of the earth was undertaken. Lord Kelvin, whose real name was William Thomson (1824–1907), was the most famous and established physicist in the United Kingdom in the second half of the nineteenth century. He was born in Glasgow, where his father was a professor of mathematics, and he studied math and physics at the University of Glasgow while still quite young. By the age of twenty he had finished his education and was already performing serious science. One of his main interests was heat transfer, which caused him to investigate the exchange of heat between the Sun and the earth. He also became curious about the propagation of heat from the center of the earth to the surface.

Kelvin had a physicist's arrogance, especially toward the other sciences, and he found the thinking of the geology community to be less than rigorous. He especially objected to Hutton and Lyell's conception of a steady-state, recycling earth, which was buttressed by unspecified chemical reactions in the core of the planet. Exhaustively studying the laws of thermodynamics, he was convinced that the Sun and the earth were cooling at a rate that was constant, though currently unknowable, in direct opposition to what the geology community believed. If the earth was cooling, then geologic processes would necessarily have been different in the past, with, for example, greater rates of volcanism and more intense winds and storms. Thus, Hutton and Lyell's belief in the constancy of geologic processes would be wrong. As early as 1844, Kelvin had decided to try to determine the age of the earth. In essence, he revisited Buffon, starting with the Frenchman's idea that the earth had begun as a mass of molten rock, and that the heat from that original molten state was slowly dissipating.

Working on other projects for years, Kelvin was finally able to announce his findings in 1862. He chose a locale guaranteed to ensure notice—the very scene of one of James Hutton's triumphs, the Royal Society of Edinburgh (which by now had moved from the university to a beautiful building in New Town). In April of that year, Kelvin, now the physics professor at the University of Glasgow, addressed the society, reading his paper "On the Secular Cooling of the Earth." The thirty-eight-year-old aggressively criticized the Huttonian view because it did not allow for his contention that the earth was cooling. After calculating a rate of cooling and then extrapolating, Kelvin presented an age of between 20 million and 400 million years. Though the speech was impressively technical, and the geologists in the audience may have felt chastised, the wide range of ages did not prove persuasive. Revisiting the issue periodically, Kelvin consistently brought his estimate down, so that by the end of the 1880s, the bottom number, 20 million, had become Kelvin's official calculation. Kelvin was such a scientific titan that 20 million years became the accepted age of the earth. Twenty million became the new 6,000—the age of the earth recognized by popular culture.

Everything changed when Henri Becquerel discovered radiation in his Paris lab in 1896. In an ingenious experiment, he took sealed, unexposed photographic film and put it in a room with uranium, a mineral known to exhibit unusual properties. When the film was checked after several hours, it was completely exposed, as if it had been left in the sunshine. Becquerel, and his lab assistants Marie and Pierre Curie, next sought to understand what had happened. They realized that certain elements are fundamentally unstable, and this instability leads their isotopes (different species of an element, the difference being the number of neutrons) to undergo spontaneous decay, to break apart, and to ultimately produce stable atoms, along with energy. This action is known as radioactive decay, and a by-product is the escape of the energy. It was this energy that caused the exposure of the film.

The discovery of radioactivity galvanized the world of science, and ambitious scientists everywhere began working in the field. The next big breakthrough occurred in 1902, when Ernst Rutherford and his colleague Frederick Soddy showed that radioactive decay of a given element occurs at a constant rate, one that can be measured. It was not long before they realized that the steady rate of radioactive decay could be used as a geologic clock to determine the age of the earth. In 1905, Rutherford delivered the Silliman Lectures at Yale, and used the forum to challenge the science community to try to date the age of the earth using this new natural clock.

A chemist soon took up the Rutherford-Soddy challenge. In 1907, Bertram Boltwood used the known rate of decay of radium, combined with his discovery that uranium decays to lead, to come up with a range of 400 million to 2,200 million years for the age of the earth.

The next push to date the earth moved back to the United Kingdom in the person of Arthur Holmes (1890–1965). Holmes was a gifted student who had won a science scholarship to study physics at Imperial College in London. When his talents were recognized by the physicist R. J. Strutt, Holmes was urged to stay on as a graduate student to work on the "age of the earth problem." Holmes followed Boltwood's insight into the relationship between uranium and lead, and came up with more refined numbers. In 1913, and again in 1927, he published a popular book on the age of the earth, in which he presented his calculation that the earth was 1.6 billion years old.

Through the 1930s and 1940s the work of Holmes, Alfred Nier, E. Gerling, and F. Houtermans became more rigorous and precise. These men, and many others, were now working primarily with common lead, and by the beginning of World War II, "isotope geologists" had now calculated the age of the earth to be at least 3.3 billion years.

The final breakthrough came in the 1950s, when Claire Patterson, of Caltech, realized that the only way to get a completely accurate measurement of common lead decay was to leave the planet, since the complicated mix of other elements in the earth distorted any measurement attempt. He and his colleagues decided to focus on objects that were the same age as all the planets in the solar system, including the earth, but allowed for more accurate lead decay calculations—meteorites. As Claire Patterson later related:

> Lead in iron meteorites was the kind of lead that was in the solar system when it was first formed, and . . . it was preserved in iron meteorites without change from uranium decay, because there is no uranium in iron meteorites. . . . If we only knew what the isotopic composition of primordial lead was in the earth when formed, we could take that number and stick it into this marvelous equation that the atomic physicists had worked out. And you could turn the crank and blip—out would come the age of the earth.

By 1956, Patterson had calculated the age of the earth to be 4.6 billion years, which remains the accepted age of our planet. James Hutton was right—the earth is unfathomably old.

WHY IS IT THAT JAMES HUTTON, the man who proved the earth's antiquity and made it possible for Claire Patterson to complete his work, is essentially unknown to all but geologists? One reason for his relative invisibility is that geology has never been a particularly flashy discipline. And it seems to have done an especially poor job of publicizing its founding fathers, whereas other scientific fields have somehow pushed their pioneers into the popular consciousness: Lavoisier in chemistry, Galileo and Newton in physics, Darwin and Gregor Mendel in biology. However, that cannot be the whole story because most people have at least heard of Charles Lyell, the discipline's other trailblazer.

Another reason is that the world's attention was certainly focused elsewhere when Hutton was first presenting and then defending his theory. The American War of Independence ended in 1783, and the French Revolution began in 1789—two galvanizing events that changed world history forever, and certainly preoccupied the people who lived through the last decades of the eighteenth century, as well as future historians. Still, these two conflagrations did not prevent Adam Smith from gaining the recognition he deserved.

One is left with the fact that James Hutton was not a gifted communicator. Indeed, just about the only negative passage in Playfair's biography concerns Hutton's writing: "The reasoning is sometimes embarrassed by the care taken to render it strictly logical; and the transitions, from the author's peculiar notions of arrangement, are often unexpected and abrupt. These defects run more or less through all Dr. Hutton's writings, and produce a degree of obscurity astonishing to those who knew him, and heard him everyday converse with no less clearness and precision than animation and force."

The defective writing, coupled with the great pain Hutton was suffering while working on his book, caused it to be put together in a hurried way. Not just long (approximately 1,200 pages over two volumes), *The Theory of the Earth* also contained turgid passages from other works in other languages. A book that unwieldy simply would not be read today, and it was not widely read then. One historian has determined that the first printing was just 500 copies (not an unusual first printing for the time; the first printing for *Origin of Species* was just 1,250 copies), but it was never reprinted. The long article from 1788 was a solid piece of scientific writing, but it was available only in a volume containing several other papers, and it was not broadly distributed.

That Hutton's book was virtually ignored by readers in 1795, and thereafter, seemed to seal his fate as a member of the legion of forgotten scientists. In fact, one might argue that the key to being remembered by posterity is to write a popular book. The works of Charles Lyell and Charles Darwin are regarded as masterpieces, still wonderfully interesting and insightful over 100 years after their publication. Adam Smith's *Wealth of Nations* carries a similar status. David Hume's books, though less widely read now, were best-sellers 200 years ago, and are actively perused by philosophy students today. Newton's *Principia,* though technical, is still read by most serious students of physics. John Playfair's own book, unlike Hutton's, was well written and popular at the time, but perhaps it was prevented from remaining visible over the decades because he was explaining another's work.

Steno, Werner, Black, and Hall also did not write books that resonated across a broad spectrum of readers, and their fates have been similar to Hutton's. But the lack of recognition for the doctor is an enormous oversight: James Hutton's brilliant insights forever changed how we think about our planet and our place on it. The man who found time has hopefully been found at last.

APPENDIX

T HE PURPOSE OF THIS SHORT appendix is threefold: to briefly locate James Hutton's ideas in the context of modern geology, to explain the current interpretation of the key sites that he used to prove his theory, and to provide definitions of the geological phenomena he viewed. For a fuller description of all of these subjects, I recommend studying Stephen Marshak's *Earth: Portrait of a Planet* (W. W. Norton, 2002), in my opinion the most complete and up-to-date text in geology.

JAMES HUTTON'S
PLACE IN MODERN GEOLOGY

James Hutton is recognized in geology circles as the "Father of Modern Geology" because his theory of the earth was the starting point for the current understanding of the workings of the dynamic planet Earth. Many historians of the field believe that Hutton's key finding was his recognition that geologic processes occurring in the world today, such as the occasional earthquake and the everyday rainstorm causing minute amounts of erosion, are no different now than in the past. There was no period of more intense volcanic activity, and no earth-covering deluges, for example. This realization is today called the principle of uniformitarianism, and geologists use the expression "The present is the key to the past" as the embodiment of this idea. Charles Lyell took uniformitarianism to a new level, and it has been one of the fundamental teachings of the profession ever since. (However, recent research has shown that there have been episodes in the earth's history when the intensity of certain phenomena, such as volcanism, has been greater than at other times.)

Uniformitarianism is a valuable insight, but its hold has been so strong that the discovery in the 1970s that the impact of an asteroid (or comet) caused the mass extinction that killed off the dinosaurs 65 million years ago was met with great resistance for years. Only the perseverance of the team of scientists headed by Walter Alvarez led to the acceptance of the impact thesis and the realization now that, yes, the present is the key to the past, but occasionally something catastrophic can occur. Thus, the current view in geology is a synthesis of Hutton and Lyell's uniformitarianism and Cuvier-inspired catastrophism. Evidence shows that there have been at least five mass extinctions since the evolution of complex life 500 million years ago, but it is not yet clear whether any of the others were caused by an impact; additional types of catastrophe may be awaiting discovery.

As important as the establishment of uniformitarianism was, even more significant was James Hutton's discovery that subterranean heat plays a role in the operation of the planet. Though other scientists had recognized that the inside of the earth was hot, it was Hutton who saw this heat as crucial to the earth's over-all workings. This was a seminal contribution because all previous investigators of the earth had turned to water as the key causal agent. Hutton's understanding of subterranean heat was the first step to the modern theory of plate tectonics. Plate tectonics is the unifying theory of geology. It stipulates that the outer, rigid layer of the earth—called the lithosphere—consists of about twenty sepa-rate plates of crust that move with respect to one another as they literally float on top of the asthenosphere, which is the hot plastic layer underneath the lithosphere. The fundamental difference between the lithosphere and asthenosphere is heat; the materials that make up the two layers are the same. All major geologic phe-nomena—volcanoes, earthquakes, and mountain formation being the most prominent—result from the interaction between the plates as they slide alongside one another, or collide, or separate.

James Hutton had no way of discerning the existence of lith-osphere plates—which were recognized beginning in the late 1960s—but his prediction that subterranean heat leads to the creation of new land by raising previously submerged sedimen-tary rocks was essentially correct. The uplift he predicted does occur, and it is caused by the interaction of plates.

THE CURRENT UNDERSTANDING
OF HUTTON'S EUREKA SPOTS

Siccar Point

Siccar Point reveals what is now called an unconformity, a sur-face at which two separate sets of rocks that were clearly formed at different times come into contact. Siccar Point was formed in the following way: The bottom formation, what Hutton called the alpine schistus, is now called Silurian graywacke (graywacke is a sandstone made up of different-sized grains). It was formed approximately 425 million years ago when colliding plates created an underwater trench in which sediments started to settle, carried in by submarine avalanches. Pressure compacted the sediment, and minerals precipitated out of water solutions to cement the grains together. Eventually, the sediment became rock, the graywacke. Over time, the movement of the plates compressed and wrinkled the sediment layers, tilting some layers into their present vertical orientation. Then this undersea formation was uplifted above sea level to form mountainous land. The uplifting was probably caused by the collision of two plates. Erosion then went to work on the new mountains.

By about 345 million years ago, the mountains of the raised Silurian graywacke had been eroded to form a plain. Parts of this plain were submerged beneath the sea. Erosion of the nearby Caledonian Mountains, a range formed by a collision of the European and North American plates, caused the deposition of sediments in river channels, floodplains, and deltas over the graywacke. These sediments eventually became the Old Red Sandstone, or Red Devonian Sandstone.

More uplifting raised the entire structure above the water again, only to be discovered by Hutton, Hall, and Playfair in 1788.

Glen Tilt

Glen Tilt is an area of igneous intrusion. Here magma rose and pushed up into preexisting rock by slowly creeping upward between grains or in cracks. The magma did not make it all the way to the surface, and thus cooled and solidified underground.

Arthur's Seat

This fascinating mound at the eastern edge of Edinburgh is the result of a complex series of events. As described in Chapter 7, the first known geologic event in the region was a period of intense volcanism starting approximately 425 million years ago. Arthur's Seat was originally a volcanic cone, probably dating from the second round of volcanism that hit the area 350 million years ago. Then the region was flooded by a sea; sediments, which later became sedimentary rocks, built up and surrounded the cone. Later earthquakes and mountain-formation pressure from colliding plates further raised and distorted the formation. Finally, an ice-age glacier overran it all. Arthur's Seat was particularly important for Hutton, though, because of the unusual wall of exposed rocks, called Salisbury Crags, that traverses the side of the mound. He alone realized that the Crags were younger rocks than the strata around them. He correctly deduced that the Salisbury Crags are what geologists today call a sill, which is an intrusion of igneous rock in between the layers of preexisting rock. This realization became important as he pondered the significance of subterranean heat.

DEFINITIONS

Breccia. Coarse sedimentary rock consisting of rock fragments; there is a layer of breccia between the Silurian graywacke and the Old Red Sandstone at Siccar Point.

Dike. A tabular (wall-shaped) intrusion of rock that cuts across the layering of preexisting rock.

Erosion. The grinding away and removal of the earth's surface materials by moving water, air, and ice.

Granite. A coarse-grained igneous rock. Werner argued that granite was the first rock precipitated from the receding universal ocean; thus it was the oldest type of rock on earth; Hutton argued that granite came from the subterranean regions of the earth and that it was often younger than sedimentary rocks (Hutton was right).

Ice Age. An interval of time in which the climate was colder than it is today and glaciers advanced to cover large areas of the continents. Mountain glaciers also grew.

Igneous Rock. Rock that forms when hot molten rock (magma or lava) cools and freezes solid.

Intrusion. Rock formed by the freezing of magma underground. The igneous veins found at Glen Tilt are intrusions.

Limestone. Sedimentary rock composed of calcite; most limestone consists of the shells of dead organisms like clams, corals, and plankton.

Lithification. The process that causes loose sediments to convert to sedimentary rocks. This transformation is caused by: (1) compaction from the weight of sediments above, and (2) cementation, caused when minerals precipitate out of water solutions passing through the sediment. Hutton was the first to understand the role of pressure, but he did not know about cementation.

Marl. A sedimentary rock that is essentially a mix of limestone and clay; it forms in coastal environments where rivers and streams flow into lakes, seas, or oceans. It is very common on the east side of England and Scotland.

Metamorphic Rock. Rock that forms when preexisting rock (either igneous or sedimentary) changes to new rock as a result of an increase in pressure and temperature. For example, marble is the metamorphic rock that results from the extreme pressuring and heating of limestone, which is a sedimentary rock. The existence of metamorphism was not understood in James Hutton's day, and many metamorphic rocks were confused with sedimentary, because both can have layering. Hutton's insistence that both subterranean heat and pressure were needed to form stratified rocks was correct for metamorphic rocks, but incorrect for sedimentary.

Plate Tectonics. The theory that the outer layer of the earth consists of separate lithosphere plates that move with respect to one another.

Precipitate. The solid crystals formed when atoms dissolved in a solution come together. Werner's system of the earth depended on precipitation for the formulation of all rocks; Werner's followers thought that rocks, such as granite, formed as the result of the receding and evaporation of the universal ocean. This is not correct; granite forms by solidification of a melt.

Rock. A coherent, naturally occurring solid, consisting of an aggregate of minerals or a mass of glass.

Sandstone. A sedimentary rock consisting almost entirely of sand grains; the sand is typically composed of quartz.

Sedimentary Rock. Rock that forms either by the cementing together of fragments broken off preexisting rock or by the precipitation of mineral crystals out of water solutions at or near the earth's surface.

Sills. A nearly horizontal (like a windowsill), tabletop-shaped tabular intrusion that injects between layers of preexisting rock.

Subsidence. The vertical sinking of the earth's surface in a region relative to a reference plane. Subsidence creates space for sediment layers to accumulate.

Unconformity. A boundary, between two different rock sequences, that represents an interval of time during which sediments were not deposited and/or were eroded.

Uplift. The vertical elevation of land. It is caused by a variety of pressures, all related to the movement of crustal plates or mantle flow. Uplift can yield mountain ranges.

Sources and Suggested Readings

ONE WORK WAS A KEY TO THIS ENTIRE project and was used in almost every chapter: John Playfair's *Biographical Account of the Late Dr. James Hutton,* also referred to as the *Life of Dr. Hutton.* This memorial was published in the fifth volume of the *Transactions of the Royal Society of Edinburgh* in 1805 (still available from the RSE Scotland Foundation), after having been read to the fellows of the Society on January 10, 1803. The *Life of Dr. Hutton* is only sixty pages long, but it is filled with invaluable information about James Hutton's life, and, perhaps most important, details about his thought processes based on conversations between Hutton and Playfair that can be gleaned from no other source.

Informing the entire book as well are the four works on Hutton's theory of the earth. These are Hutton's brief abstract from 1785, which was never published (though it was typeset), and which carried the title "Abstract of a Dissertation Concerning the System of the Earth, Its Duration, and Stability"; the article from the *Transactions of the Royal Society of Edinburgh,* officially titled "Theory of the Earth; or an Investigation of the Laws Observable in the Composition, Dissolution, and Restoration of Land upon the Globe," which appeared in 1788 and was the document written by Hutton that created the biggest stir among other scientists; Hutton's two-volume *Theory of the Earth, with Proofs and Illustrations* (Edinburgh: William Creech, 1795); and John Playfair's *Illustrations of the Huttonian Theory of the Earth* (Edinburgh: William Creach, 1802). In 1970, Hafner Publishing Company published a volume that compiled the abstract, the 1788 paper, and Playfair's memorial.

Several secondary sources by the leading experts on James Hutton were very useful, especially for the second half of the book: Dennis R. Dean's *James Hutton and the History of Geology* (Ithaca: Cornell University Press, 1992) and his "James Hutton and His Public, 1785–1802," *Annals of Science,* vol. 30 (1973); Jean Jones's "James Hutton," in *The Scottish Enlightenment, 1730–1790: A Hotbed of Genius,* ed. David Daiches, Peter Jones, and Jean Jones (Edinburgh: The Saltire Society, 1996) and her "James Hutton's Agricultural Research and His Life as a Farmer," *Annals of Science,* vol. 42 (1985); and Donald R. McIntyre and Alan McKirdy's *James Hutton: The Founder of Modern Geology* (Edinburgh: The National Museums of Scotland Publishing, 2001). Each of these fine scholars has written numerous articles on specific topics, and these are cited below.

Finally, I found myself, time and again, consulting the magisterial *Dictionary of Scientific Biography,* ed. Charles Coulston

Gillispie (New York: Charles Scribner's Sons, 1970–1978), to acquire essential information about the numerous, sometimes obscure, scientists referred to in the book.

What follows below is a list of the chief sources utilized while researching each chapter.

PROLOGUE

David Daiches, "The Scottish Enlightenment," in David Daiches, Peter Jones, and Jean Jones, eds., *The Scottish Enlightenment, 1730–1790: A Hotbed of Genius* (Edinburgh: The Saltire Society, 1996).

Adrian Desmond and James Moore, *Darwin: The Life of a Tormented Evolutionist* (London: Penguin Books, 1991).

Robert B. Downs, *Books That Changed the World* (London: Penguin Books, 1956).

Jack Finegan, *Handbook of Biblical Chronology* (Princeton, NJ: Princeton University Press, 1964).

Jean Jones, "James Hutton," in David Daiches, Peter Jones, and Jean Jones, eds., *The Scottish Enlightenment, 1730–1790: A Hotbed of Genius* (Edinburgh: The Saltire Society, 1996).

John Playfair, "Life of Dr. Hutton," in *Transactions of the Royal Society of Edinburgh,* vol. 5 (Edinburgh: Royal Society of Edinburgh, 1805).

CHAPTER ONE

Gordon Y. Craig, "Siccar Point: Hutton's Classic Unconformity," in *Scottish Borders Geology,* ed. A. D. McAdam, E. N. K. Clarkson, and P. Stone (Edinburgh: Scottish Academic Press, 1993).

Gordon Y. Craig, Donald B. McIntyre, and Charles D. Waterston (with Jean Jones and others), *James Hutton's Theory of the Earth: The Lost Drawings* (Edinburgh: Scottish Academic Press, 1978).

V. A. Eyles, "Sir James Hall," in *Dictionary of Scientific Biography,* ed. Charles Coulston Gillispie (New York: Charles Scribner's Sons, 1970–1978).

V. A. Eyles, "James Hutton," in *Dictionary of Scientific Biography,* ed. Charles Coulston Gillispie (New York: Charles Scribner's Sons, 1970–1978).

Jean Jones, "James Hutton," in *The Scottish Enlightenment, 1730–1790: A Hotbed of Genius,* ed. David Daiches, Peter Jones, and Jean Jones (Edinburgh: The Saltire Society, 1996).

David Land, "What Else Did Hutton See at Siccar Point," *Edinburgh Geology,* vol. 21 (1989).

Stephen Marshak, *Earth: Portrait of a Planet* (New York: W. W. Norton, 2002).

James G. Playfair, "Biographical Account of the Late Professor Playfair," in *The Works of John Playfair* (Edinburgh: A. Constable & Co., 1822).

John Playfair, "Life of Dr. Hutton," in *Transactions of the Royal Society of Edinburgh,* vol. 5 (Edinburgh: Royal Society of Edinburgh, 1805).

CHAPTER TWO

F. J. Bacchus, "The Chronicle of Eusebius," *The Catholic Encyclopedia* (New York: Robert Appleton Company, 1910).

F. J. Bacchus, "Eusebius of Caesarea," *The Catholic Encyclopedia* (New York: Robert Appleton Company, 1910).

F. J. Bacchus, "Theophilus," *The Catholic Encyclopedia* (New York: Robert Appleton Company, 1910).

Timothy David Barnes, *Constantine and Eusebius* (Cambridge, MA: Harvard University Press, 1981).

James Barr, "Why the World Was Created in 4004 B.C.: Archbishop Ussher and Biblical Chronology," *The John Rylands University Library Journal,* vol. 67 (1985).

G. Brent Dalrymple, *The Age of the Earth* (Stanford, CA: Stanford University Press, 1991).

Jack Finegan, *Handbook of Biblical Chronology* (Princeton, NJ: Princeton University Press, 1964).

Adrian Fortescue, "Julius Africanus," *The Catholic Encyclopedia* (New York: Robert Appleton Company, 1910).

Charles Coulston Gillispie, *Genesis and Geology* (Cambridge, MA: Harvard University Press, 1951).

Colin Groves, "From Ussher to Slusher, from Archbishop to Gish: Or, Not in a Million Years . . . ," *Archeology in Oceania,* vol. 31 (1996).

A. Vander Heeren, "The Septuagint Version," *The Catholic Encyclopedia* (New York: Robert Appleton Company, 1910).

H. Leclercq, "The First Council of Nicea," *The Catholic Encyclopedia* (New York: Robert Appleton Company, 1910).

Bruce M. Metzger and Michael D. Coogan, eds., *The Oxford Companion to the Bible* (New York and Oxford: Oxford University Press, 1993).

Stephen Toulmin and June Goodfield, *The Discovery of Time* (New York: Harper & Row, 1965).

CHAPTER THREE

W. J. Baird, *The Scenery of Scotland, the Structure Beneath* (Edinburgh: National Museums of Scotland, n.d.).

Walter Biggar Blaikie, *Edinburgh at the Time of the Occupation* (Edinburgh: T. and A. Constable, 1910).

Gale E. Christianson, *In the Presence of the Creator: Isaac Newton and His Times* (New York: The Free Press, 1984).

I. Bernard Cohen and Richard S. Westfall, *Newton: A Norton Critical Edition* (New York: W. W. Norton, 1995).

David Daiches, *Edinburgh* (London: Hamish Hamilton, 1978).

David Daiches, *James Boswell & His World* (New York: Scribner, 1976).

T. M. Devine, *The Scottish Nation, 1700–2000* (London: Penguin Books, 1999).

Peter and Fiona Somerset Fry, *The History of Scotland* (London: Routledge, 1982).

Nathaniel Harris, *Heritage of Scotland* (New York: Checkmark Books, 2000).

Arthur Herman, *How the Scots Invented the Modern World* (New York: Crown Publishers, 2001).

Jean Jones, "James Hutton," in *The Scottish Enlightenment, 1730–1790: A Hotbed of Genius,* ed. David Daiches, Peter Jones, and Jean Jones (Edinburgh: The Saltire Society, 1996).

David McAdam, *Edinburgh: A Landscape Fashioned by Geology* (Edinburgh: The British Geological Survey, 1993).

John Playfair, "Life of Dr. Hutton," in *Transactions of the Royal Society of Edinburgh,* vol. 5 (Edinburgh: Royal Society of Edinburgh, 1805).

J. F. Scott, "Colin Maclaurin," in *Dictionary of Scientific Biography,*
Charles Coulston Gillispie, ed. (New York: Charles Scribner's
Sons, 1970–1978).

Robert Louis Stevenson, *Edinburgh* (London: Seeley, Service & Co., 1912).

Richard S. Westfall, *The Life of Isaac Newton* (Cambridge: Cambridge
University Press, 1993).

CHAPTER FOUR

Walter Biggar Blaikie, *Edinburgh at the Time of the Occupation*
(Edinburgh: T. and A. Constable, 1910).

Robert Chambers, *History of the Rebellion in Scotland, in 1745, 1746*
(Philadelphia: E. C. Mielke, 1833).

David Daiches, *Charles Edward Stuart: The Life and Times of Bonnie
Prince Charlie* (London: Thames and Hudson, 1973).

Arthur Herman, *How the Scots Invented the Modern World* (New York:
Crown Publishers, 2001).

CHAPTER FIVE

Frank Dawson Adams, *The Birth and Development of the Geological
Sciences* (London: Constable and Company, 1938).

Peter Bowler, *The Earth Encompassed* (New York: W. W. Norton, 1993).

Albert Carozzi, "Benoit de Maillet," in *Dictionary of Scientific
Biography,* ed. Charles Coulston Gillispie (New York: Charles
Scribner's Sons, 1970–1978).

A. C. Crombie, "René du Perron Descartes," in *Dictionary of Scientific
Biography,* ed. Charles Coulston Gillispie (New York: Charles
Scribner's Sons, 1970–1978).

G. Brent Dalrymple, *The Age of the Earth* (Stanford, CA: Stanford
University Press, 1991).

Archibald Geikie, *The Founders of Geology* (London: Macmillan, 1897).

Stephen Jay Gould, "Hutton's Purpose," in *Hen's Teeth and Horse's Toes*
(New York: W. W. Norton, 1983).

Stephen Jay Gould, *Time's Arrow, Time's Cycle* (Cambridge: Harvard
University Press, 1987).

Stephen Jay Gould, "The Titular Bishop of Titiopolis," in *Hen's Teeth
and Horse's Toes* (New York: W. W. Norton, 1983).

Suzanne Kelly, "Thomas Burnet," in *Dictionary of Scientific Biography*, ed. Charles Coulston Gillispie (New York: Charles Scribner's Sons, 1970–1978).

Charles Lyell, *Principles of Geology*, vol. 1 (London: J. Murray, 1830; reprint, Chicago: University of Chicago Press, 1990).

John Playfair, "Life of Dr. Hutton," in *Transactions of the Royal Society of Edinburgh*, vol. 5 (Edinburgh: Royal Society of Edinburgh, 1805).

R. Rappaport, "Guillaume-François Rouelle," in *Dictionary of Scientific Biography*, ed. Charles Coulston Gillispie (New York: Charles Scribner's Sons, 1970–1978).

Jacque Roger, "George-Louis Buffon," in *Dictionary of Scientific Biography*, ed. Charles Coulston Gillispie (New York: Charles Scribner's Sons, 1970–1978).

Keith S. Thomson, "Vestiges of James Hutton," *American Scientist*, vol. 89 (May–June 2001).

Stephen Toulmin and June Goodfield, *The Discovery of Time* (New York: Harper & Row, 1965).

Richard S. Westfall, "Robert Hooke," in *Dictionary of Scientific Biography*, ed. Charles Coulston Gillispie (New York: Charles Scribner's Sons, 1970–1978).

CHAPTER SIX

Lawrence Airey, "A Brief History of Berwick" (distributed by the Berwick Upon Tweed Civic Society, 1989).

Jean Jones, "James Hutton's Agricultural Research and His Life as a Farmer," *Annals of Science*, vol. 42 (1985).

E. C. Mossner, *The Life of David Hume*, 2d ed. (Oxford: Oxford University Press, 1980).

John Playfair, "Life of Dr. Hutton," in *Transactions of the Royal Society of Edinburgh*, vol. 5 (Edinburgh: Royal Society of Edinburgh, 1805).

CHAPTER SEVEN

R. G. W. Anderson, "Joseph Black," in *The Scottish Enlightenment, 1730–1790: A Hotbed of Genius*, ed. David Daiches, Peter Jones, and Jean Jones (Edinburgh: The Saltire Society, 1996).

Alexander Broadie, ed., *The Scottish Enlightenment: An Anthology* (Edinburgh: Canongate Books, 1997).

Norman E. Butcher, "James Hutton's House at the St. John's Hill, Edinburgh," in *Book of the Old Edinburgh Club* (Edinburgh: T. and A. Constable, 1997).

N. Campbell, R. Martin, and S. Smellie, *The Royal Society of Edinburgh* (Edinburgh: The Society, 1983).

Archibald Clow and Nan L. Clow, "Dr. James Hutton and the Manufacture of Sal Ammoniac," *Nature,* vol. 159 (1947).

David Daiches, "The Scottish Enlightenment," in *The Scottish Enlightenment, 1730–1790: A Hotbed of Genius,* ed. David Daiches, Peter Jones, and Jean Jones (Edinburgh: The Saltire Society, 1996).

Harold Dorn, "James Watt," in *Dictionary of Scientific Biography,* ed. Charles Coulston Gillispie (New York: Charles Scribner's Sons, 1970–1978).

V. A. Eyles and Joan M. Eyles, "Some Geological Correspondence of James Hutton," in *Annals of Science,* vol. 7 (1951).

Adam Ferguson, "Life of Dr. Black," in *Transactions of the Royal Society of Edinburgh,* vol. 5 (Edinburgh: Royal Society of Edinburgh, 1805).

Henry Guerlac, "Joseph Black," in *Dictionary of Scientific Biography,* ed. Charles Coulston Gillispie (New York: Charles Scribner's Sons, 1970–1978).

Arthur Herman, *How the Scots Invented the Modern World* (New York: Crown Publishers, 2001).

Jean Jones, "The Geological Collection of James Hutton," *Annals of Science,* vol. 41 (1984).

Jean Jones, "James Hutton and the Forth and Clyde Canal," *Annals of Science,* vol. 39 (1982).

Jean Jones, Hugh S. Torrens, and Eric Robinson, "The Correspondence Between James Hutton and James Watt, Part 1," *Annals of Science,* vol. 51 (1994).

Jean Jones, Hugh S. Torrens, and Eric Robinson, "The Correspondence Between James Hutton and James Watt, Part 2," *Annals of Science,* vol. 52 (1995).

Peter Jones, "David Hume," in *The Scottish Enlightenment, 1730–1790: A Hotbed of Genius,* ed. David Daiches, Peter Jones, and Jean Jones (Edinburgh: The Saltire Society, 1996).

E. C. Mossner, *The Life of David Hume* (Oxford: Oxford University Press, 1980).

John Playfair, "Life of Dr. Hutton," in *Transactions of the Royal Society at Edinburgh*, vol. 5.

John Rae, *Life of Adam Smith* (London: Macmillan & Co., 1895).

D. D. Raphael, "Adam Smith," in *The Scottish Enlightenment, 1730–1790: A Hotbed of Genius*, ed. David Daiches, Peter Jones, and Jean Jones (Edinburgh: The Saltire Society, 1996).

Eric Robinson and Douglas McKie, *Partners in Science* (London: Constable, 1970).

CHAPTER EIGHT

Gordon Y. Craig, Donald B. McIntyre, and Charles D. Waterston (with Jean Jones and others), *James Hutton's Theory of the Earth: The Lost Drawings* (Edinburgh: Scottish Academic Press, 1978).

Dennis R. Dean, "James Hutton and His Public, 1785–1802," *Annals of Science*, vol. 30 (1973).

Dennis R. Dean, *James Hutton and the History of Geology* (Ithaca, NY: Cornell University Press, 1992).

Dennis R. Dean, "James Hutton on Religion and Geology: The Unpublished Preface to His Theory of the Earth," *Annals of Science*, vol. 32 (1975).

Dennis R. Dean, ed., *James Hutton in the Field and in the Study* (Delmar, NY: Scholar's Facsimiles & Reprints, 1997).

M. D. Eddy, "Geology, Mineralogy and Time in John Walker's University of Edinburgh Natural History Lectures, 1779–1803," *History of Science*, vol. 39 (2001).

James Hutton, "Abstract of a Dissertation Concerning the System of the Earth, Its Duration, and Stability," in *James Hutton's System of the Earth* (Darien, CT: Hafner Publishing Co., 1970).

James Hutton, "Observations on Granite," in *James Hutton's System of the Earth* (Darien, CT: Hafner Publishing Co., 1970).

James Hutton, "Theory of the Earth; Or an Investigation of the Laws Observable in the Composition, Dissolution, and Restoration of Land upon the Globe," in *James Hutton's System of the Earth* (Darien, CT: Hafner Publishing Co., 1970).

James Hutton, *Theory of the Earth, with Proofs and Illustrations* (Edinburgh: William Creech, 1795).

Donald B. McIntyre and Alan McKirdy, *James Hutton: The Founder of Modern Geology* (Edinburgh: National Museums of Scotland Publishing Limited, 2001).

John Playfair, *Illustrations of the Huttonian Theory of the Earth* (Edinburgh: William Creech, 1802).

John Playfair, "Life of Dr. Hutton," in *Transactions of the Royal Society of Edinburgh,* vol. 5 (Edinburgh: Royal Society of Edinburgh, 1805).

E. L. Scott, "Richard Kirwan," in *Dictionary of Scientific Biography,* ed. Charles Coulston Gillispie (New York: Charles Scribner's Sons, 1970–1978).

Harold W. Scott, "John Walker," in *Dictionary of Scientific Biography,* ed. Charles Coulston Gillispie (New York: Charles Scribner's Sons, 1970–1978).

CHAPTER NINE

Robert P. Beckinsale, "Jean Andre De Luc," in *Dictionary of Scientific Biography,* ed. Charles Coulston Gillispie (New York: Charles Scribner's Sons, 1970–1978).

Franck Bourdier, "Georges Cuvier," in *Dictionary of Scientific Biography,* ed. Charles Coulston Gillispie (New York: Charles Scribner's Sons, 1970–1978).

Gordon Y. Craig, Donald B. McIntyre, and Charles D. Waterston (with Jean Jones and others), *James Hutton's Theory of the Earth: The Lost Drawings* (Edinburgh: Scottish Academic Press, 1978).

Georges Cuvier, *Essay on the Theory of the Earth,* 2d ed. (Edinburgh: William Blackwood, 1815).

Joan M. Eyles, "Robert Jameson," in *Dictionary of Scientific Biography,* ed. Charles Coulston Gillispie (New York: Charles Scribner's Sons, 1970–1978).

V. A. Eyles, "Sir James Hall," in *Dictionary of Scientific Biography,* ed. Charles Coulston Gillispie (New York: Charles Scribner's Sons, 1970–1978).

Derek Flinn, "James Hutton and Robert Jameson," *Scottish Journal of Geology,* vol. 16 (1980).

James G. Playfair, "Biographical Account of the Late Professor Playfair," in *The Works of John Playfair* (Edinburgh: A. Constable & Co., 1822).

Martin J. S. Rudwick, "Adolphe-Theodore Brongniart," in *Dictionary of Scientific Biography,* ed. Charles Coulston Gillispie (New York: Charles Scribner's Sons, 1970–1978).

Martin J. S. Rudwick, "Louis-Constant Prevost," in *Dictionary of Scientific Biography,* ed. Charles Coulston Gillispie (New York: Charles Scribner's Sons, 1970–1978).

CHAPTER TEN

Philip Appleman, *Darwin,* Norton Critical Edition, 3d ed. (New York: W. W. Norton, 2001).

John Bowlby, *Charles Darwin: A New Life* (New York: W. W. Norton, 1991).

Walter F. Cannon, "William Buckland," in *Dictionary of Scientific Biography,* ed. Charles Coulston Gillispie (New York: Charles Scribner's Sons, 1970–1978).

Charles Darwin, *The Autobiography of Charles Darwin, 1809–1882,* ed. Nora Barlow (London: Collins, 1958).

Charles Darwin, *The Origin of Species* (London: John Murray, 1859).

Adrian Desmond and James Moore, *Darwin: The Life of a Tormented Evolutionist* (London: Penguin Books, 1991).

Charles Lyell, *Life, Letters, and Journals of Sir Charles Lyell* (London: John Murray, 1881).

Charles Lyell, *Principles of Geology,* vol. 1 (London: J. Murray, 1830; reprint, Chicago: University of Chicago Press, 1990).

Mark Ridley, *The Darwin Reader,* 2d ed. (New York: W. W. Norton, 1996).

Leonard G. Wilson, *Charles Lyell, the Years to 1841: The Revolution in Geology* (New Haven, CT: Yale University Press, 1972).

Leonard G. Wilson, "Lyell: The Man and His Times," in *Lyell: The Past Is the Key to the Present,* ed. D. J. Blundell and A. C. Scott (London: Geological Society, 1998).

Leonard G. Wilson, "The Origins of Charles Lyell's Uniformitarianism," in *Uniformity and Simplicity,* ed. Claude C. Albritton (The Geological Society of America, 1967).

EPILOGUE

Jed Z. Buchwald, "William Thomson/Lord Kelvin," in *Dictionary of Scientific Biography*, ed. Charles Coulston Gillispie (New York: Charles Scribner's Sons, 1970–1978).

G. Brent Dalrymple, *The Age of the Earth* (Stanford, CA: Stanford University Press, 1991).

Cherry Lewis, *The Dating Game: One Man's Search for the Age of the Earth* (Cambridge: Cambridge University Press, 2000).

Alfred Romer, "Antoine-Henri Becquerel," in *Dictionary of Scientific Biography*, ed. Charles Coulston Gillispie (New York: Charles Scribner's Sons, 1970–1978).

Chris Stassen, "Chronology of Radiometric Dating," unpublished article, 1998.

OTHER RECOMMENDED READINGS

Frank Dawson Adams, *The Birth and Development of the Geological Sciences* (London: Constable and Company, 1938).

Claude Allegre, *From Stone to Star: A View of Modern Geology* (Cambridge, MA: Harvard University Press, 1992).

Walter Alvarez, *T. Rex and the Crater of Doom* (Princeton, NJ: Princeton University Press, 1997).

Stephen J. Gould, *Time's Arrow, Time's Cycle: Myth and Metaphor in the Discovery of Geological Time* (Cambridge, MA: Harvard University Press, 1987).

Rachel Laudan, *From Mineralogy to Geology: The Foundations of a Science, 1650–1830* (Chicago: University of Chicago Press, 1987).

John A. McPhee, *Annals of the Former World* (New York: Farrar, Straus & Giroux, 1998).

E. K. Peters, *No Stone Unturned: Reasoning About Rocks and Fossils* (New York: W. H. Freeman and Company, 1996).

Martin J. S. Rudwick, *Scenes from Deep Time* (Chicago: University of Chicago Press, 1992).

Jenny Uglow, *The Lunar Men* (New York: Farrar, Straus & Giroux, 2002).

Simon Winchester, *The Map That Changed the World: William Smith and the Birth of Modern Geology* (New York: HarperCollins, 2001).

Acknowledgments

I HAVE BEEN A BOOK EDITOR FOR twenty years, but until working on my own enterprise, I never truly appreciated how collaborative the book-writing process is. Quite simply, I could not have written this study without a tremendous amount of assistance from a number of key people who were experts on James Hutton, Scotland, geology, history, books, or some combination of the above. A small group went beyond the call of duty and were of fundamental importance to the project. I would like to thank them here.

Gary Hincks, the most accomplished geology artist I have ever worked with, was tremendously supportive. Sending books, making sketches, helping me visualize scenes, providing introductions, and otherwise lending me his broad understanding about geology and the United Kingdom, his early active aid was essential for getting the book off the ground.

Jean Jones, an outstanding historian of science and among the small group of scholars who have made substantial contributions to the extant literature on James Hutton, has been enormously generous with her time and expertise. In addition to directing me to all her own relevant articles as well as those of others, each of which was extensively used, she read large parts of the manuscript and saved me from many embarrassing mistakes, misstatements, and misleading comments. Of course, any remaining errors are entirely my own.

Stephen Marshak, of the University of Illinois, is a gifted geologist and the author of one of the leading introductory textbooks in the field. He has consistently been available to "talk geology" with me. Steve took an active interest in the Hutton book, and with his guidance, I have tried mightily to present the geology correctly.

Peter Nevraumont, of Nevraumont Publishing Company, served as a much-needed sounding board for me in the early stages of the project, and his broad reading and incredible good sense helped me devise how I was going to tell the story of James Hutton.

Dr. Michael A. Taylor, Curator of Vertebrate Paleontology at the National Museums of Scotland, offered tremendous assistance during my visit to Edinburgh, providing me with many useful publications and sources. His staff members went out of their way to be supportive and courteous.

Leonard G. Wilson, of the University of Minnesota, is among the leading scholars on Charles Lyell and early nineteenth-century geology. He was very generous with his knowledge, and was a tremendous resource for the final chapters.

Two families enriched my visit to Scotland: the Marshalls, who own James Hutton's old farm, Slighhouses; and the Drysdales, who own the land that includes Siccar Point. Both were extraordinarily pleasant and cooperative.

In addition to these individuals, various other people have given their time and talent, and I would like to take this opportunity to recognize them. First, the book would have been impossible to write without the important research and spade-work done by the small group of historians of science who have pieced together the details of the underdocumented life of James Hutton. Their research is impressive, and they are all excellent writers as well: Dennis R. Dean, the already mentioned Jean Jones, and Donald B. McIntyre. Similarly, Leonard G. Wilson's rich and insightful research into Charles Lyell was invaluable.

I first learned about James Hutton while working on the geology books that I published at Princeton University Press and W. W. Norton & Company in the 1990s, and I would like to thank all my geology authors and friends who stimulated me to appreciate their fascinating, challenging, and dynamic field: Rick Alley, Walter Alvarez, Bob Ballard, Tony Dahlen, Dick Fisher, Grant Heiken, Paul Hoffman, Dick Holland, Andy Knoll, David Lamb, the already mentioned Steve Marshak, George Philander, Wylie Poag, Chris Scholz, Bill Schopf, Dan Schrag, Art Snoke, Jeroen Tromp, Ben van der Pluijm, and Cindy van Dover.

The Man Who Found Time is a work of popular history, too, and I was fortunate to have some wonderful teachers who turned me on to history many years ago. They have no idea how much they have enriched my life: the late, great Kent Forster; Robert Green, Charles Hunnel, Steve Kale, Michael MacDonald, Bill Tomey, and especially Jim Donnelly and Jack Spielvogel.

Many friends and colleagues either read the proposal or chapters or discussed ideas at length with me, and they were of great service as I sorted through the many issues related to the story. In particular, I would like to express my appreciation to Mandy Brown, Peter Dougherty, George Funkhouser, Susan Gaustad, Charles Gillispie, Roby Harrington, Richard Hughes, Marian Johnson, Don Lamm, Drake McFeely, Kurt Radke, Pam Scholer, Janet Stites, Wynn Sullivan, and Angela von der Lippe. Leelo Marjamaa was instrumental during the final execution of the proposal.

I have been blessed during my career in working with scores of outstanding scholars and writers. All my authors have taught me something new about a subject or about writing in general, and I thank them all. A few have been the lynchpins of my career and I would like to recognize them here: Padma Desai, Bill Greene, Bennett McCallum, and especially Dave Kreps.

My agent, Susan Rabiner, an author herself and talented former editor, was simply outstanding during the tough conceptual stages of the endeavor. I am still amazed by how much work she invested in my effort and how much good sense she applied to it. She is known in the publishing industry as among the best at what she does, and now I know why.

I thank all the folks at Perseus Publishing who have been so patient, committed, and supportive, in particular David Goehring, Carolyn Savarese, and Elizabeth Carduff, three of the top pros in the book business. Thanks also to Erica Lawrence and Jennifer Blakebrough-Raeburn for their careful attention through the copyediting stage.

As important as everyone mentioned above has been, if this book has any popular appeal, it is because of my editor, Amanda Cook. Every interaction I had with her, and there were many, taught me something new about the editorial craft. Whoever started the rumor in the publishing industry that editors do not edit anymore has not encountered Amanda—she worked tirelessly on my manuscript. She was motivating and smart, filled with good ideas, and, most important, she knew where to focus and where to let go. I feel the book is as much hers as mine.

Thanks also to my great kids, Julie, Christie, Karl, Kerry, Laura, and now Clay, too, who gave up weekends with Dad for a couple of years; and to my wonderful wife, Donna, for coming up with the book's title, for reading every word of the manuscript, and for giving me a number of good suggestions.

Finally, I would like to thank my parents, Claire and Jack Repcheck, to whom this book is dedicated, for all their backing over the years, and for raising me and my brothers and sisters in a household that celebrated curiosity.

When I started on this book, I had hoped it would be an intellectual adventure, but my overriding memory will always be of a real adventure. Seeing Siccar Point with my parents and Donna on a rainy June day in 2002 from the fields just below it was remarkable enough—it is such a spectacular sight—but then being chased out of that wet pasture by twenty angry cows really seared the memory. I now know why James Hutton chose to explore the area by boat.

INDEX

Aberdeen University, 59
Act of Union with England, 56–57,
	65
Adam, Robert, 128
Alaric, 38
Alexander the Great, 30
Alvarez, Walter, 210
American War of Independence,
	205
Analytical Review, 161
Angles, 48
*Annales Veteris Testamenti (Annals of
	the Old Testament)* (Ussher),
	42
Apocalyptic prophesy, 39. *See also*
	Biblical chronology
Arduino, Giovanni, 148–149
Arran, 158, 164, 167–168
Arthur's Seat, 64, 125–126, 213
Atheism, 5
Athol, Duke of, 156
Attila the Hun, 38–39
Augustine, St., 5–6
Auld Reekie, 52
Auvergne, 186

Bakewell, Robert, 180
	Introduction to Geology, 180
Bank of Scotland, 56
Basalt, 171. *See also* Subterranean
	heat, existence of
Becquerel, Henri, 202
Bede the Venerable, 5–6
	biblical chronology of, 40, 42
Bell, John, 109
Berwick, 105–106
Bible
	and Luther, Martin, 41
	translation of, and Jerome, St.,
		29–30
	See also individual bibles
Biblical chronology, 1–6, 5–6, 14,
	94–95, 153
	of Bede the Venerable, 40, 42
	of Eusebius, 28–31, 41
	of Fiore, Joachim, 40
	of Isidore of Seville, 40, 42
	of Jerome, St., 37–38
	and Jesus Christ, second coming
		of, 35–37, 39–40, 41
	of Josephus, Flavius, 31–32

of Julius Africanus, 32–37, 32(n3),
    41, 42, 43
of Kepler, Johannes, 43
of Luther, Martin, 40–41, 42, 43
of Newton, Sir Isaac, 42–43,
    97–98
and recalculation, 37–38
of Theophilus of Antioch, 32(n3)
of Ussher, James, 42–43
See also Earth, age of; Earth,
    theory of
Biblical geology, 4, 97
and Lyell's Principles of Geology,
    199
Black, Joseph, 9, 119, 129, 130–134,
    135, 136, 138, 140–141, 142,
    143, 153, 160, 163–164, 167,
    172–173, 190, 206–207
"Experiments on Magnesia Alba,
    Quicklime, and other Alcaline
    Substances," 132
and Hutton's theory of earth,
    presentation of, 150–151
Blair, 156
Boar Club, 140
Boerhaave, Hermann, 87, 90
Boltwood, Bertram
and earth, age of, calculation of,
    203
Bonnie Prince Charlie. See Stuart,
    Charles Edward
Book of Daniel, 43
Book of Genesis, 1, 3, 4–5, 6, 31, 94.
    See also Biblical chronology
Book of Moses, 31
Book of Revelation, 6, 43
Borders, 104–107
Brahe, Tycho, 61
Brongniart, Alexandre, 174

Buckland, William, 180–182, 183,
    184, 191
and earth, theory of, 181, 185
and Lyell's Principles of Geology,
    199
Buffon, G. L. de, 99–100, 101,
    101(n1), 153
and earth, theory of, 148
Histoire Naturelle, 99–100, 148
Bulgars, 39
Bunkle Castle, 107, 115
Burgundians, 38
Burnet, Thomas, 97–98
The Sacred Theory of the Earth,
    97–98
Burnett, Elspeth, 88
Burns, Robert, 128
Burt, Edward, 52–54
Byzantine Empire, 39

Calculus, invention of, 61
Caltech, 204
Cambridge University, 6, 191
Cameron clan
and Jacobite Uprising of 1745,
    71–82
Cape Club, 139–140
Carbon dioxide, discovery of,
    131–132, 142, 172
Carlisle, 79
Castle Rock, 46–47, 48, 50
Catastrophism, 175, 183–184, 191,
    192, 210
and Lyell's Principles of Geology,
    199
Catholic Church, 14
Chile, 185
Chirnside, 106
Christian Bible, 30

Christian church
and earth, age of, 13–14
Christianity
and Constantine I, 26–27, 37–38
Chronologia (Julius), 33–37
Chronology (Eusebius), 29–30, 30–31, 33, 37–38
Clerk, John, 128, 155, 164, 167
An Essay on Naval Tactics, 128
Clerk-Maxwell, George, 109, 112–113, 123, 136, 155, 156–157
Color, properties of, 61
Commission for the Forfeited Annexed Estates, 112
"Concerning the System of the Earth, Its Duration, and Stability" (James Hutton), 151
Constantine I, 26–27, 26(n1)
and Christianity, 37–38
and Eusebius, 28
Constantinople, 26(n1)
Cope, Jonathan, 73–74, 76, 77, 78
Copernicus, Nicolaus, 2–3, 5, 7, 61
De Revolutionibus Oribium Coelestium, 3
Corriearrick, 74
Cosmogonies, 97–101
Council of Nicaea, 26–28, 40
Crag and tail, 47
Crichton-Browne, Sir James, 124
Critical Review, 160
Crochallan Fencilbles, 140
Crop rotation, 111
Cullen, William, 119, 128, 132, 143, 145, 147, 153
Culloden Moor, battle of, 80–81, 86
Cumberland, Duke of, 81–82, 118
Curie, Marie, 202

Curie, Pierre, 202
Custom House, 138
Cuvier, Georges, 178, 183–184
and earth, theory of, 174–175, 176, 181, 185
Cycles, notion of, 90–91

D'Alembert, Jean, 121
Damascus, Pope, 29
Darien Affair, 55–56
Dark Ages, 38
Darwin, Charles, 5, 6–7, 11, 103, 129, 153, 188–198, 205, 206
and earth, age of, calculation of, 200
and evolution, theory of, insight for, 191–193. *See also* Evolution, theory of
and Jameson, Robert, 189–191, 192
and Lyell, Charles, 191–193, 196, 197–198
Origin of Species, 196, 197–198, 200, 206
and Sedgwick, Adam, 191
Darwin, Erasmus, 129, 189
David, King, 48
Davie, James, 91–92, 93
Davie/Hutton Sal Ammoniac chemical works, 92, 123–124, 125
De' crostacei (Moro), 97
Deism, 5, 62–63
De Luc, Jean André, 167, 170, 171
and earth, theory of, 165, 174
Deluge, 18
and Lyell's *Principles of Geology,* 199
See also Noah's Flood; Universal Ocean theory

De Revolutionibus Oribium
     Coelestium (Copernicus), 3
De sanguine et circulatione in
     microcosmo (The Blood and the
     Circulation of the Microcosm)
     (James Hutton), 90–91
Descartes, René, 61, 95
Dialogue Concerning the Two Chief
     World Systems (Galileo), 3
Diderdot, Denis, 121
Dinner parties, and intellectual
     exchange, 139
Discourse of Earthquakes (Hooke), 96
Dissertation Concerning a Sold Body
     Enclosed By the Process of
     Nature Within a Solid (Steno),
     95–96
Drainage ditches, 110
Drummond, George, 74–75, 85–86,
     86–87, 117–118, 119–120
Duke of Solm's ironworks, 147
Dunglass Burn beach, 19, 23
Duns, 106–107
Duns Castle, 107
Dybold, John, 93

Earth, age of, 1–11
     calculation of, 199–204
     and Christian church, 13–14
     and Lyell, Charles, 6–7
     See also Biblical chronology;
          Earth, theory of
Earth, as center of universe, 2–3
Earth: Portrait of a Planet (Marshak),
     209
Earth, theory of
     and Buckland, William, 181,
          185
     and Buffon, G. L. de, 148

and Cuvier, Georges, 174–175,
     176, 181, 185
and De Luc, Jean André, 165, 174
and Hutton, James, 4–5, 6–8,
     10–11, 14–15, 17–18,
     114–116, 209–211
and key books, c. 1752, 95–101
and Werner, Abraham Gottlob,
     147–150, 153, 181, 185. *See*
     also Universal ocean theory
See also Biblical chronology;
     Earth, age of
Earth processes, 95–96, 101
Earthquakes, 96
Edict of Milan, 27
Edinburgh, 9
     building structures in, 51–52
     geologic history of, 45–47
     and Jacobites, purging of, 117–118
     landmarks in, 46–50
     and New Town, 119–120
     plumbing system in, 52–54
     social structure in, 52
     tenements in, 52
     weather in, 50–51
Edinburgh High School, 58–59
Edinburgh (R. Stevenson), 50–51
Edwin, King (of Northumbria), 48
Elements of Geognosy (Jameson), 173
Elijah, 5–6, 36, 41
The Encyclopedia Britannica (ed.
     Smellie), 128, 167, 176–177
England
     march on, and Jacobite Uprising of
          1745, 78–92, 85
     and Scotland, union with, 56–57,
          65
England, southwest, 135–136
English Anglican Church, 14

Erosion, dynamics of, 8, 19, 64, 65, 110, 113–114, 115–116, 134, 210. *See also* Huttonian theory

Erratics, 176

An Essay on Naval Tactics (Clerk), 128

Essay on the History of Civil Society (Ferguson), 128

Essay on the Natural History of the Earth (James Hutton), 113–114

Essay on the Principle of Population (Mathus), 195–196

Essay Toward a Natural History of the Earth (Woodward), 98

Euclidian geometry, 16

Eusebius, 27–28, 39
  biblical chronology of, 28–31, 41
  Chronology, 29–30, 30–31, 33, 37–38
  and Julius's *Chronologia,* 37

Evolution, theory of, 5, 7, 11
  insight for, 191–193
  See also Darwin, Charles

Experimental geology, 172

"Experiments on Magnesia Alba, Quicklime, and other Alcaline Substances" (Black), 132

Eyemouth, 106

Father of Modern Geology, 209

Ferguson, Adam, 128, 132, 143, 154
  Essay on the History of Civil Society, 128

Fiore, Joachim
  biblical chronology of, 40

Firth of Clyde, 122

Firth of Forth, 122

Fixed air, 131

Flodden Field, battle of, 49

Flodden Wall, 49–50, 51

Forth and Clyde Canal, construction of, 122–123

Forth and Clyde Canal committee, 123

Forth and Clyde Navigation Company, 122–123

The 45. *See* Jacobite Uprising of 1745

France
  and Stuart, Charles Edward, 69–70, 78–79

Franklin, Benjamin, 121, 129

Franks, 38, 39

Freiburg School of Mines, 147–148

French Enlightenment, 90

French Revolution, 89, 174, 205

Galapagos Islands, 193–194

Galileo, 3, 5, 7, 61, 95, 205
  Dialogue Concerning the Two Chief World Systems, 3

Galloway, 157–158

Geological Society, 174, 180, 182, 191

Geology, 6–8

George II, 72, 81

Gerling, E., 203

Glaciers, 176

Glasgow, 80, 122, 123

Glasgow University, 59

Glenfinnan, 71

Glen Tilt, 156–157, 164, 184, 212–213

Gododdins, 48

Gould, John, 194

Gould, Stephen Jay, 7–8, 114

Grampian Mountains, 156

Granite, 155–158. *See also*
  Subterranean heat, existence of

Great Glen Fault, 113

Greyfriar's Cemetery, 163
Gutenberg, Johannes, 30
Gutted Haddie, 64

Hall, James, 11, 94, 129, 160, 164,
      167–168, 175–176, 178–179,
      185, 206–207
   and Huttonian theory, defense of,
      170–173
   and North Sea coast, 16–23
   "On the Revolutions of the Earth's
      Surface," 176
Hall, John, 93, 94
Halley, Edmund, 97
Hardy, Thomas, 104
Harrying of the glens, 81–82
Heat and pressure, insights into, 132,
      133–134, 134(n1). *See also*
      Subterranean heat, existence of
Hebrew Bible, 30–31
Highlander clans
   and Jacobite Uprising of 1745,
      67–82
Highlands, 112–113
Highland Scotland, 68
Histoire Naturelle (Buffon), 99–100,
      148
The History of England (Hume), 121
History of Scotland (Robertson), 128
HMS *Beagle*, 6
Holmes, Arthur
   and earth, age of, calculation of,
      203, 204
Holyrood Abbey, 48–49
Hooke, Robert, 96
   Discourse of Earthquakes, 96
Hooker, Joseph, 197
Hope, Thomas, 190
Houtermans, F., 203

Hume, David, 9, 63, 88, 120–122,
      129, 130, 132, 136, 139, 206
   The History of England, 121
   A Treatise of Human Nature,
      120–121
Hume, Joseph, 88–89
Hutton, Isabella, 163
Hutton, James
   in Arran, 158
   at Arthur's Seat, 64, 125–126
   birth of, 43, 45
   "Concerning the System of the
      Earth, Its Duration, and
      Stability," 151
   daily habits, 126–127
   death of, 160–161, 163
   *De sanguine et circulatione in
      microcosmo (The Blood and the
      Circulation of the Microcosm),*
      90–91
   early life, 54–64
   and earth, age of, calculation of,
      199–200
   and earth, theory of, 4–5, 6–8,
      10–11, 14–15, 17–18,
      114–116, 209–211. *See also*
      Huttonian theory
   and earth, theory of, book on, 161
   and earth, theory of, critics of,
      154–155, 164–167, 173
   and earth, theory of, paper on, 159
   and earth, theory of, paper on,
      review of, 159–161
   and earth, theory of, presentation
      of, 145–146, 150–154
   at Edinburgh High School, 58–59
   Essay on the Natural History of
      the Earth, 113–114
   estate of, 163–164

and family farm. *See* Slighhouses
family holdings, 108. *See also*
    Slighhouses
and farming, 93–94, 103–116
    in Galloway, 157–158
    in Glen Tilt, 156–157
and Highlands, tour of, 112–113
and illegitimate son, 88–89, 91,
    124, 164
and law apprenticeship, 64
in London, 91
and medical school, 64, 65
at Norfolk, 93–94
and North Sea coast, 13–24
personality of, 9–10
recognition for, lack of, 204–207
siblings of, 124
at Siccar Point, 21–22, 158, 170
in southwest England and Wales,
    135–136
and St. John's Hill, house at,
    124–125
The Theory of the Earth, 161, 164,
    168, 206
at University of Edinburgh, 59–63
at University of Leyden, 90–91
at University of Paris, 88–90
Hutton, John, 107
Hutton, Sarah Balfour, 43, 54–55, 58
Hutton, William, 43, 54, 55, 56, 69,
    107, 108
    death of, 57–58
Huttonian theory, 190
and Hall, James, defense of,
    170–173
and Lyell, Charles, defense of,
    182–189, 190, 192
and Lyell's *Principles of Geology,*
    199

and Playfair, John, defense of,
    167–170, 173
vs. Wernerian theory, 17–18,
    166–167, 173, 183. *See also*
    Wernerian theory
See also Hutton, James, and earth,
    theory of

Ice Age, 8, 176
*Illustrations of the Huttonian Theory
    of the Earth* (Playfair),
    168–170, 173, 185, 206
Impact thesis, 210
Imperial College, 203
Industrial Revolution, 9
Inquisition, 3
Introduction to Geology (Bakewell), 180
Isidore of Seville
    biblical chronology of, 40, 42
Isle of Isk, 187
Isle of Wight, 184
Isotope geologists, 203
"Is the Huttonian Theory of the Earth
    Consistent with Fact?"
    (Jameson), 165–166
Isthmus of Darien, 55
Italy, 39, 186–187
Iznit, Turkey, 26(n1)

Jacobites, 57
    purging of, in Edinburgh, 117–118
Jacobite Uprising of 1689, 68–69
Jacobite Uprising of 1715, 69
Jacobite Uprising of 1745, 67–82, 117
    and battle method, 76–77
    and Culloden Moor, battle of,
        80–81, 86
    and England, march on, 78–92, 85
    and Preston, battle of, 76–78

James, Robert, 175
James II, 55, 57, 65, 68–69
James III, 68, 69
Jameson, Robert, 165–167, 168, 170,
    171, 173, 176, 178
    and Darwin, Charles, 189–191, 192
    Elements of Geognosy, 173
    "Is the Huttonian Theory of the
        Earth Consistent with Fact?",
        165–166
    and Lyell's *Principles of Geology,*
        199
    Mineralogy of Scotland, 166
    *An Outline of the Mineralogy of the
        Shetland Islands, and of the
        Island of Arran,* 166
Jardin du Roi, 90, 99–100
Jefferson, Thomas, 121
Jerome, St.
    and Bible, translation of, 29–30
    biblical chronology of, 37–38
    and Eusebius's *Chronology,* 37–38
    and Eusebius's *Chronology,*
        translation of, 29–30
Jesus Christ, 6, 27–28
    birth of, and Eusebius, 28
    birth of, and Julius's *Chronologia,*
        35
    second coming of, and biblical
        chronology, 35–37, 39–40, 41
Jewish Antiquities (Josephus), 32–33
Jewish Talmud, 36
Josephus, Flavius, 30, 31–32, 39
    biblical chronology of, 31–32
    Jewish Antiquities, 32–33
Julius Africanus, 30, 39
    biblical chronology of, 32–37,
        32(n3), 41, 42, 43
    Chronologia, 33–37

Kant, Immanuel, 128
Keeper of the Advocates Library, 121
Kelvin, Lord. *See* Thomson, William
Kepler, Johannes, 61
    biblical chronology of, 43
King James Bible, 3, 30, 42
Kinnordy, 184–185
Kirwan, Richard, 160–161, 164–165,
    166, 167, 170, 171, 174
    *Transactions of the Royal Irish
        Academy,* 160–161
Knox, John, 58

Lammermuir Hills, 108
Lavoisier, Antoine, 16, 90, 205
Lawnmarket, 58
Lead decay calculations, 203–204
Lehmann, Johann, 149
Leibnitz, Gottfried, 61, 99, 101
    Protoggea, 99
Lexicon Technicum, 63
Libya, 33
Life of Dr. Hutton (Playfair),
    113–114, 205
Limestone, 172–173. *See also*
    Magnesia alba; Subterranean
    heat, existence of
Lincoln's Inn, 182
Linnean Society, 182, 197
Lithification, 134(n1)
Lithosphere plates, 211
Lochs Bakie, 184
London, 91
London Times, 194
Louis XIV, 69, 89
Lowlands, 57
Luther, Martin, 5–6
    biblical chronology of, 40–41, 42, 43
    Supputatio Annorum Mundi, 41

Lutheran Church, 14
Lyell, Charles, 11, 179–180, 205, 206,
    210
    and Darwin, Charles, 6–7,
        191–193, 196, 197–198
    and earth, age of, calculation of,
        200
    and Huttonian theory, defense of,
        182–189, 190, 192
    Principles of Geology, 6–7,
        187–188, 191, 196, 199

Macbeth, 48
Macdonald clan
    and Jacobite Uprising of 1745,
        67–82
Maclaurin, Colin, 60–61, 62–63, 65,
    74–75, 91, 117, 118, 119,
    141–142
    *Sir Isaac Newton's Philosophical
        Discoveries,* 61
    Treatise of Fluxions, 60 61
Maclaurin, John, 143
Magnesia alba, 131–132, 142. *See also*
    Limestone; Subterranean heat,
    existence of
Maillet, Benoit de, 99, 101
    Telliamed, 99
Malcolm III, 48
Mantell, Gidean, 182–183, 184, 185
Marshak, Stephen
    Earth: Portrait of a Planet, 209
Mary II, 57, 68
Mathus, Thomas, 195–196
    *Essay on the Principle of
        Population,* 195–196
McGowan, Mr., 141
Mechanics, science of, mathematical
    synthesis of, 61

*Memoir on the Geology of Central
    France* (Scrope), 186
Mendel, Gregor, 205
Merse, 104
Meteorites, and lead decay
    calculations, 204
Mineralizing principle, for
    sedimentary rock, 133–134,
    134(n1). *See also* Sedimentary
    rock
Mineralogy of Scotland (Jameson), 166
Mirror Club, 140
Moderate Party, 118
Monro, Alexander, 87, 132, 142
Monthly Review, 161
Moro, Anton-Lazzaro, 96–97
    De' crostacei, 97
Mossner, E. C., 88–89
Murray, George, 72–73, 76, 78, 79–80
Murray, John, 197
Museum of Natural History, 147, 166
Muslims, 39

*The Natural History of the Mineral
    Kingdom* (Williams), 160
Natural philosopher, vs. scientist,
    14(n1)
Nether Monynut, 108
Newcomen steam engine, 135
New Theory of the Earth (Whiston),
    98
Newton, Sir Isaac, 4, 14, 16, 60–62,
    65, 119, 205
    biblical chronology of, 42–43,
        97–98
    and Boerhaave, Hermann, 87
    and cycles, notion of, 90–91
    Opticks, 62
    Principia, 61, 206

Newtonian Revolution, 61
New Town, 119–120
Nicaea, 26, 26(n1)
Nier, Alfred, 203
Noah's Flood, 4, 8, 95, 97–98, 99. *See also* Deluge; Universal ocean theory
Norfolk, 93–94
Nor' Loch, 120
North Sea coast, 13–24
Numerology, 36

Oldham, William, 96
Old Testament, 30–31, 43
    and Julius's *Chronologia,* 33–37
    See also individual books
Olympiad system, 29
"On the Revolutions of the Earth's Surface" (Hall), 176
"On the Secular Cooling of the Earth" (Thomson), 201
Opticks (Newton), 62
Origin of Species (Darwin), 196, 197–198, 200, 206
Otto of Freising, 40
An Outline of the Mineralogy of the Shetland Islands, and of the Island of Arran (Jameson), 166
Oxford University, 180
Oyster Club, 140–141

Palace of Holyroodhouse, 49, 76
Panama, 55–56
Panmure House, 138
Paris, 174, 183–184
Paterson, William, 55–56
Patterson, Claire
    and earth, age of, calculation of, 204
Pease Bay, 20, 21

Pentateuch, 31
Philosophical Society, 141–143
Pitt, William (the younger), 137
Plate tectonics, 8, 211
Playfair, James, 177
Playfair, John, 10, 11, 64, 88, 93–94, 126, 129, 141, 145, 153, 156–157, 160, 161, 164, 176, 185
    death of, 177
    and Huttonian theory of earth, defense of, 167–170, 173
    Illustrations of the Huttonian Theory of the Earth, 168–170, 173, 185, 206
    Life of Dr. Hutton, 113–114, 205
    and North Sea coast, 15–24
Poker Club, 140
Presbyterian Church, 118–119
Presbyterian Reformation, 58, 59
Preston, battle of, 76–78
Prevost, Constant, 183–184
Primary micaceous schistus, 19
Principia (Newton), 61, 206
Principles of Geology (Lyell), 6–7, 187–188, 191, 196, 199
Pringle, Sir John, 16
Printing press
    and Luther, Martin, 41
Proceedings of the Royal Society, 159
Protestant Reformation, 40
Protoggea (Leibnitz), 99
Ptolemy, 3
Ptolemy II, 30–31

Quarterly Review, 186

Radioactive decay calculations, 202–203

Radioactivity, discovery of, 202
Radium decay calculations, 203
Ramsay, Allan, 139
Rankenian Club, 140
Recalculation
    and biblical chronology, 37–38
Reed Point, 20
Renton, John, 107
Robertson, William, 118, 128, 143, 145
    History of Scotland, 128
Robison, John, 134, 143, 145
Rock groups
    classification scheme for, 148–149
    See also Granite; Sedimentary rock
Roman Empire, fall of, 38–39
Rouelle, Guillaume-François, 90
Rousseau, Jean-Jacques, 121
Royal Irish Academy, 160
Royal Medical Society, 165
Royal Society, 14, 16, 96, 141–142,
    142–143, 145, 201
    Physical Class and Literary Class,
    146
Russel, Mr., 161
Rutherford, Ernst, 202–203
Rutherford, John, 88

The Sacred Theory of the Earth
    (Burnet), 97–98
Sal ammoniac, production of, from
    coal soot, 91–92
Salisbury Crags, 126, 190, 213
Saussure, Horace de, 154
Scientist, vs. natural philosopher,
    14(n1)
Scotch Board of Customs, 137–138
Scotland
    and England, union with, 56–57,
    65

Scots plough, 110
Scott, Sir Walter, 128–129
Scottish Enlightenment, 9, 11,
    127–130
    and intellectual exchange,
    139–143
    participants in, 128–130
Scottish Presbyterian Church, 14
Scrope, George, 186
    *Memoir on the Geology of Central
    France,* 186
Secondary sandstone strata, 19. *See
    also* Subterranean heat,
    existence of
Sedgwick, Adam, 191, 192
Sedimentary rock, 8, 96, 113–114,
    115–116, 133–134, 134(n1).
    See also Limestone;
    Subterranean heat, existence of
Select Society, 140
Septuagint Bible, 42
    and Eusebius's *Chronology,*
    30–31
    and Josephus's *Antiquities,* 32
    and Julius's *Chronologia,* 34–35
Siccar Point, 21–22, 158, 164, 170,
    179–180, 185, 211–212
Silliman Lectures, 203
Silurian graywacke, 19
*Sir Isaac Newton's Philosophical
    Discoveries* (Maclaurin), 61
Slighhouses, 93, 103–116, 124
    and cattle and sheep, raising of,
    111
    farm workers at, 111–112
    grounds, 107
Smellie, William, 128
    The Encyclopedia Britannica (ed.),
    128, 167, 176–177

Smith, Adam, 9, 129, 132, 136–139, 140–141, 143, 145, 153, 205, 206
  The Theory of Moral Sentiments, 136
  Wealth of Nations, 129, 136, 137–138, 206
Social clubs, and intellectual exchange, 139–143
Society for the Improvement of Medical Knowledge, 142
Soddy, Frederick, 202–203
Sorbonne, 100
Spain
  and Panama, 55–56
St. Andrews University, 59
St. Giles Cathedral, 49
St. Jago, 6–7, 191–193
St. John's Hill, 124–125
Steam engine, two-cylinder, separate condenser, 135
Steno, Nicolaus, 95–96, 101, 206–207
  Dissertation Concerning a Sold Body Enclosed By the Process of Nature Within a Solid, 95–96
Stevenson, John, 63
Stevenson, Robert Louis, 50–51
  Edinburgh, 50–51
Stewart, Archibald, 85–86
Stewart, Matthew, 119, 168
Stirling Castle, 85
Strutt, R. J., 203
Stuart, Charles Edward, 65, 67–83, 85, 117
  and France, 69–70, 78–79
Subterranean heat, existence of, 8, 18–19, 22, 170–173, 210–211. *See also* Huttonian theory

Suffolk plough, 110
Supputatio Annorum Mundi (Luther), 41
Sussex, 182–183, 184
Swediaur, Dr., 141
Sylvester, Pope, 27(n2)

Telescope, invention of, 61
Telliamed (Maillet), 99
Theophilus of Antioch biblical chronology of, 32(n3)
The Theory of Moral Sentiments (Smith), 136
The Theory of the Earth (James Hutton), 161, 164, 168, 206
Thomas Aquinas, St., 3, 5–6
Thomson, William
  and earth, age of, calculation of, 200–202
  "On the Secular Cooling of the Earth," 201
Transactions of the Royal Irish Academy (Kirwan), 160–161
Treatise of Fluxions (Maclaurin), 60–61
A Treatise of Human Nature (Hume), 120–121
Tuillibardine, Marquis of, 72
Tweed River, 107

Uniformitarianism, 209–210
Universal gravitation, discovery of, 61–62
Universal ocean theory, 17–18, 99, 100
  and Werner, Abraham Gottlob, 17–18, 101(n1), 149, 155, 174–175, 181, 184
  See also Deluge; Noah's Flood

Universe
  age of, 2
  earth as center of, 2–3
  See also Biblical chronology;
    Earth, age of
University of Edinburgh, 15, 59–63,
    189–190
University of Edinburgh medical
    school, 74–75, 86–88, 119
University of Glasgow, 119, 201
University of Leyden, 87, 90–91
University of Paris, 88–90
Upper Old Red Sandstone, 19
Ussher, James, 36
  *Annales Veteris Testamenti
    (Annals of the Old Testament)*
    (Ussher), 42
  biblical chronology of, 42–43

Vandals, 38–39, 39
Vespasian, Emperor, 32
Visigoths, 38, 39
Volcanoes, 97, 185–186
Vulgate, 30

Wales, 135–136
Walker, John, 142, 146–147, 154,
    156, 166–167
Wallace, Alfred Russel, 7, 196–197, 198
War of the Austrian Succession, 69
Watt, James, 9, 129, 132, 134–135,
    135, 167
Weald Valley, 200

Wealth of Nations (Smith), 129, 136,
    137–138, 206
Werner, Abraham Gottlob, 170, 171,
    206–207
  and earth, theory of, 147–150,
    153, 181, 185
  and universal ocean theory, 17–18,
    101(n1), 149, 155, 174–175,
    181, 184
  See also Wernerian theory
Wernerian theory, 190–191
  vs. Huttonian theory, 17–18,
    166–167, 173, 183. *See also*
    Huttonian theory
  See also Werner, Abraham Gottlob
Whigs, 56–57, 65, 82–83, 118–119,
    127
Whiston, William, 98
  New Theory of the Earth, 98
White Adder, 107
William III, 57, 68
Williams, John, 160
  *The Natural History of the Mineral
    Kingdom*, 160
Woodward, John, 98
  *Essay Toward a Natural History of
    the Earth*, 98

Yale University, 203
Young Pretender. *See* Stuart, Charles
    Edward

Zoological Society, 194